SpringerBriefs in Ethics

Springer Briefs in Ethics envisions a series of short publications in areas such as business ethics, bioethics, science and engineering ethics, food and agricultural ethics, environmental ethics, human rights and the like. The intention is to present concise summaries of cutting-edge research and practical applications across a wide spectrum.

Springer Briefs in Ethics are seen as complementing monographs and journal articles with compact volumes of 50 to 125 pages, covering a wide range of content from professional to academic. Typical topics might include:

- Timely reports on state-of-the art analytical techniques
- A bridge between new research results, as published in journal articles, and a contextual literature review
- A snapshot of a hot or emerging topic
- In-depth case studies or clinical examples
- Presentations of core concepts that students must understand in order to make independent contributions

More information about this series at http://www.springer.com/series/10184

Vojin Rakić

The Ultimate Enhancement of Morality

 Springer

Vojin Rakić
Center for the Study of Bioethics
University of Belgrade
Belgrade, Serbia

ISSN 2211-8101 ISSN 2211-811X (electronic)
SpringerBriefs in Ethics
ISBN 978-3-030-72472-6 ISBN 978-3-030-72473-3 (eBook)
https://doi.org/10.1007/978-3-030-72473-3

© The Author(s), under exclusive license to Springer Nature Switzerland AG 2021
This work is subject to copyright. All rights are solely and exclusively licensed by the Publisher, whether the whole or part of the material is concerned, specifically the rights of translation, reprinting, reuse of illustrations, recitation, broadcasting, reproduction on microfilms or in any other physical way, and transmission or information storage and retrieval, electronic adaptation, computer software, or by similar or dissimilar methodology now known or hereafter developed.
The use of general descriptive names, registered names, trademarks, service marks, etc. in this publication does not imply, even in the absence of a specific statement, that such names are exempt from the relevant protective laws and regulations and therefore free for general use.
The publisher, the authors and the editors are safe to assume that the advice and information in this book are believed to be true and accurate at the date of publication. Neither the publisher nor the authors or the editors give a warranty, expressed or implied, with respect to the material contained herein or for any errors or omissions that may have been made. The publisher remains neutral with regard to jurisdictional claims in published maps and institutional affiliations.

This Springer imprint is published by the registered company Springer Nature Switzerland AG
The registered company address is: Gewerbestrasse 11, 6330 Cham, Switzerland

I dedicate this book to Tea, Sara, Andrija and Filip

Preface

One of the purposes of my professional life is to write philosophical/(bio)ethical books that are, at least to a significant extent, also understandable to those readers who are not philosophers/bioethicists. I am aware that my scientific articles are no exception in having a rather narrow readership. My books have a somewhat different purpose: to broaden the readership to a certain extent, without compromising the quality of my arguments. One of the ways I do that is not to enter profound polemics on the specificities of certain issues that require substantial expertise, but to address essential themes an interdisciplinary audience may or may not be interested to deepen/broaden himself/herself on the basis of referenced literature.

Part of this approach is to write books that hardly exceed 100 pages. Readers will be able to read such books without losing momentum, obtain an insight into various issues that are essential for (bio)ethics/philosophy, become familiar with the solutions others and I propose and have the opportunity to deepen my arguments with the literature that may confirm, disconfirm or give new insights into the elaborations.

This book is no exception. It deals on a limited number of pages with the themes of Good and Evil—*in an interdisciplinary manner*. The theory behind it is largely based on the issue of the "comprehension-motivation gap".

- The comprehension-motivation gap is the phenomenon that humans do not behave as they believe they *ought* to behave (in a moral sense).

The book also argues that being good and being happy are in a circularly supportive relationship[1]. Consequently, moral enhancement, including moral bioenhancement (MBE), should have the maximization of happiness as its foundational rationale. This happiness should be grounded on the happiness most people derive most of the time by moral acting. It should not be based on the lowering of the probability of existential harms, or "ultimate harm", as has been formulated in the writings of Ingmar Persson and Julian Savulescu[2].

---

[1] Pulcu, Zahn and Elliott (2013) try to establish in their opinion paper whether guilt and unhappiness (sadness) are in a similar relationship. This book will not broaden the argument in that direction.

[2] For example, in Persson and Savulescu (2011).

In Part I of this book, certain for our purposes relevant issues in the history of thought on happiness and morality will be briefly reviewed. A new ethical concept will be proposed, one that is related to the circularly supportive relationship between morality and happiness *and* to the possible inevitability of the emergence of posthumans. It will be argued that existing humans have a moral duty to support the eventual emergence of a morally superior posthuman species. This species will not replace existing humans or their immediate descendants. It will rather evolve with the evolution of morality. It may also emerge *de novo*, without gradually replacing existing humans. In none of those cases will the happiness of existing humans be brought into question, as they will not suffer harm. On the contrary, their participation in the development of a morally superior posthuman species might only bring them joy: they will not be sacrificed but gradually surrounded by those who are morally superior. This species might bring the human species only good, precisely because it is morally superior.

- The activities of existing humans that contribute to more moral and happy lives, as well as to the eventuation of a morally superior species, I will call *Ultimate Morality (UM)*.

These arguments will be expanded in Part II of the book. Different types of Evil will be discussed in the first section. Throughout Part II of the book, I use the Appendix of the *Abolition of Man*, authored by the Christian apologetic C. S. Lewis, as a useful listing of moral rules that appear to have trans-cultural and trans-historical validity. The issue of how humans can address Evil will be discussed. It will be shown that moral education can help in cases in which some individuals do not *understand* why something is good or bad. MBE might help when certain humans know what is good and what is evil in a given situation, but don't have the motivation to act in line with this knowledge. In cases in which there is no broad social consensus on certain moral issues, public deliberation is frequently the best way to proceed. The only group of people who do not make any distinction between the morally right and morally wrong I define as psychopaths. In their case, both moral education and MBE are very difficult to realize. I will devote special sections to issues concerning sexual (im)morality and political (im)morality, as both figure quite prominently in contemporary ethical discourse.

In the second section of Part II, a number of foundations of morality will be discussed ("the Good"). It will be argued that part of our morality is rooted in certain moral intuitions that we apparently have been born with. In that context, empathy, retribution and fortitude will be addressed. The value of fortitude will be reflected through the lens of morality in politics.

In the third section of Part II, it will be argued that some of the foundations of the trans-culturally and trans-historically accepted moral rules may be subjected to revision. The ethics of UM offers such a revision in at least one essential aspect of human moral functioning that it advocates: UM, practiced in a *global state* inhabited and run by *posthumans*, is the morally most desirable outcome of the development of humankind. This outcome does not deny the validity of the trans-geographically and trans-historically accepted moral notions that will be cited *in extenso* and discussed because of their universal authority, but it does raise the possibility of rethinking the groundwork of these notions in a global state of posthumans.

The reasons for my recurrent references to the thoughts of C. S. Lewis are largely that he argued that certain moral values are shared by humankind in general. I agree, but advocate a revision of the groundwork of these values in a future inhabited by posthumans. My frequent citations of the common human values are aimed at showing that humanity shares certain foundations of morality. The common values are largely valid today and there is no reason to revise them. I am very far from advocating a re-evaluation of the primary moral values of humanity. The numerous citations of these values show why such a re-evaluation would be misplaced. But their *groundwork* cannot remain the same with the emergence of a new posthuman species. The final sections of this book elaborate on the emergence of such a species and advocate a new moral groundwork that is to accompany the eventuation of posthumans.

Belgrade, Serbia   Vojin Rakić

References

Erdem, P., Roland, Z. and Rebecca E. 2013. "The Role of Self-Blaming Moral Emotions in Major Depression and Their Impact on Social-Economical Decision Making". Front Psychol. 4: 310.
Persson, I., and Julian, S. 2011. *Unfit for the Future*. Oxford: Oxford University Press.

Acknowledgements

I wish to thank various scholars and friends for the invaluable discussions we had about numerous arguments raised in this book and for the support they gave to its publication. They include, but are certainly not limited to, John Harris, Ingmar Persson, Julian Savulescu, Nicholas Agar, Arthur Caplan, Robert Sparrow, Peter Singer, Anders Sandberg, Milan Ćirković, Katrien Devolder, Amnon Carmi, Harris Wiseman, Thomas Douglas, Bert Gordijn, Thomasine Kushner, Erik Parens, Oliver Feeney, Sarah Chan, Josephine Johnston, Aleksandar Damjanović, Yves Agid, Nada Gligorov, Shai Linn and Russell D'Souza.

For their administrative and logistical support I am primarily indebted to Milica Milašinović, Ana Stanković, Dejan Pejović, Danilo Polić, and Stefan Mićić.

My gratitude goes to Floor Oosting for her superb handling of the publication process, as well as to Christopher Wilby for his important role in it.

I am extremely grateful to my family for their patience and tolerance they had for me during the many years of my devotion to the themes I address in this book. Without them I would not have been able to write this book and the many articles that preceded it.

Praise for *The Ultimate Enhancement of Morality*

"Vojin Rakić is a bioethicist who is always worth reading and continues to produce works on important topics. Here he writes about one of the most challenging issues of our times, namely the possibility of using science and technology to permanently change humanity itself. This change Vojin believes will take place at the level of morality. Vojin Rakić argues that the next major step in our evolution will be the development of a (new?) species he believes will be *morally* superior to existing humans. It will, Vojin argues, be guided by "Ultimate Morality", a concept that will continue to have certain foundations in traditional ethics."

—John Harris

"Rakić's book is jammed with refreshing and daring insights about human enhancement. Anyone who's interested in philosophical debate about the future of humanity should read it."

—Nicholas Agar

Contents

Part I Morality and Happiness, Psychopathy and Utilitarianism

1 **Morality: Innate, Universal?** 3
   Reference 8

2 **Morality: Origin and Future** 9
   Reference 12

3 **Free Will and Moral Bioenhancement as a Compulsory Means
   of Avoiding "Ultimate Harm"** 13
   References 18

4 **Past and Present of Happiness and Morality** 21
   References 28

5 **The Future of Happiness and Morality, Psychopathy
   and Utilitarianism** 31
   5.1 The Future of Happiness and Morality 31
   5.2 Psychopathy 32
   5.3 Utilitarianism and Its Redefinition as the Maximization
       of Median Happiness 34
   References 36

6 **The Morality of a Global State, the Existence of Morally Superior
   Beings and "Special Suicide"—Brief Conceptual Explanations** 39
   6.1 The Morality of a Global State 39
   6.2 The Existence of Morally Superior Beings 40
   6.3 Introducing Special Suicide 41

Part II Evil and Good

7 Evil 45
  7.1 Deprivation of the Greatest Good 46
  7.2 Deprivation of Lesser Goods 50
  7.3 Is "Evil" Always Evil?: Sexual Adultery, Prostitution, Sugaring,
      Paraphilias 53
  7.4 The Evil Minding of Others' Business: Reality TV as a Platform
      for Mass Gossip that Is Morally Inferior to Pornography 63
  References 64

8 Good 1 65
  8.1 Empathy 65
  8.2 Justice, Rertribution 66
  8.3 Evolutionary Morality—Did It Make Us Good or Evil? 67
  8.4 Fortitude and Politics 69
  8.5 Ethics and Politics—Science Might Make Politics More Moral 70
  References 73

9 Good 2 75
  9.1 Making Moral Enhancement Work 75
  9.2 A Morally Superior Posthuman Species Guided by Ultimate
      Morality 77
  9.3 Conceptual Clarifications 77
  9.4 Ultimate Morality and the Human Emancipation
      from the Survival-at-Any-Cost Bias 78
      9.4.1 Inductive Argument—Probability 78
      9.4.2 Deductive Argument—Proof 79
  9.5 Posthuman Morality: Morality Ultimate,
      Comprehension-Motivation Gap Superseded 80
  9.6 Kant 81
  9.7 A World State of the Future—Again 82
  References 84

Part I
Morality and Happiness, Psychopathy and Utilitarianism

Chapter 1
Morality: Innate, Universal?

> *"If nothing is self-evident, nothing can be proved. Similarly if nothing is obligatory for its own sake, nothing is obligatory at all"*, C.S. Lewis, The Aboliton of Man

Why to be good? Answers to this question vary. Religious people will be inclined to invoke Divine commands, deontologists the duty to act in a way that can be universalized as a moral rule, utilitarians the maximization of happiness as the aspired outcome, other consequentialists different outcomes, virtue ethicists the improvement of moral character, contractarians obligations derived from a contract. But what if someone invokes the argument that we should act merely in our narrow self-interest, without any moral scruples—as long as we can get away with it? "Getting away with it" implies that the perpetrator of immoral actions does not suffer any adverse consequences because of them, including institutional punishment, regret, shame,[1] contempt or social exclusion. In that case, we have no reason to be good, unless it is in our self-interest. Consequently, morality becomes superfluous. Ethics loses its *raison d'etre*. The answer to the question "Why to be good" is then "There is absolutely no reason to be good, unless we have a self-interest in being good".

Is there a way to counter this type of moral nihilism by showing that there *is* reason to be good? As there is no rational proof that we ought to be moral and not to guide ourselves merely by self-interest, the only way of showing that there *is* reason to be good is to define self-interest as a moral good. That is, among else, what I am going to do in this book. I will show that what we intuitively consider as moral is conducive to our happiness. Moreover, happiness turns out to be conducive to the behavior we (intuitively) consider as moral. *Hence, our self-interest (happiness) and moral behavior are in a circularly supportive relationship.* This relationship is supported

---

[1] If we don't behave in a way we think we *ought* to behave we may feel regret, to a higher or lower degree. We may also feel shame. Regret is a step in the direction of bridging the comprehension-motivation gap, because it is part of our selves. In regret we are aware of the is-ought discrepancy in ourselves and are sometimes able to motivate ourselves to bridge it. Shame is not part of our selves in such a way. We largely feel shame because of others. In shame, our motivation to change, *because we wish to become better people for its own sake*, is missing in the way it is present in regret.

by a variety of relevant experiments that have been conducted to this effect. They will be discussed in this book. If this relationship can be proven, we have strong reasons to be good. These reasons are rooted in our self-interest. Hence, it will be argued that the answer to the question "Why to be good" is that all alternatives to goodness are worse, including the attempt to pursue self-interest by immoral behavior.

If we conclude that we have reason to be good, the question arises where we derive the sense of goodness from. How did we develop a sense of morality? Did we develop it or have we been born with a moral apparatus? Is morality a social construct, is its basis genetic, did it develop through evolution, did God endow us with morality, or is morality a combination of some of these factors?

Various findings show that our sense of morality is not a social construct. As the research of Paul Bloom, author of *Just Babies*, demonstrates, babies are endowed with a sense of morality (Bloom 2013). Similar findings in the case of animals have been presented by Frans de Waal. One important conclusion of research in this area is not that humans are born good, but that they are born with a sense of morality. They are not *born* good because they sometimes or frequently fail to act in line with their sense of morality. Both human babies and animals can be aggressive and show a lack of empathy toward out-groups. But a sense of morality, empathy, altruism and goodness human babies and certain animals do appear to have. If this sense is not manifested towards outgroups, it tends to be manifested toward ingroups.[2]

There is an abundance of evidence for this, evidence that is not limited to humans. Primatologist Frans de Waal writes in *The Bonobo and the Atheist* (2013) that even rats show empathy. Faced with a choice between two containers, one with chocolate chips and another with a trapped companion, rats often choose to rescue their companions first. In addition, Macaque monkeys, more distant from humans on the evolutionary chain than the great apes, will not take food if this causes another monkey harm.[3]

Human babies are capable of not only telling the difference between right and wrong, but also of making morally relevant decisions. It is noteworthy that they develop this capability before they learn to speak and even hold up their bodies.

Paul Bloom argues that babies who are too young to have learned about morality, have an innate moral sense. Moreover, they show a basic disposition to goodness. For example, infants start sharing after they are six months old. Toddlers will already help a stranger in need.

There is an abundance of evidence suggesting that babies have a sense of morality:

1. In one study by Felix Warneken and Michael Tomasello, a toddler was in a room with his mother when a stranger walked in with his hands full. The stranger walked over to a closet to open the door but couldn't manage it. As this drama

---

[2]See https://www.theatlantic.com/health/archive/2013/11/as-babies-we-knew-morality/281567/; https://www.psychologytoday.com/us/blog/moral-landscapes/201701/is-humanitys-moral-sense-inherited-or-nurtured; last accessed on 1 November, 2020.

[3]For the paragraphs that follow, see https://www.theatlantic.com/health/archive/2013/11/as-babies-we-knew-morality/281567/; https://www.psychologytoday.com/us/blog/moral-landscapes/201701/is-humanitys-moral-sense-inherited-or-nurtured); last accessed on 1 November, 2020.

was unfolding, no one looked at the toddler or encouraged him to do anything. Yet about half of all of the infants tested spontaneously got up and walked over to the closet to open the door for the person in need—an all the more remarkable feat when you realize that toddlers are very reluctant to approach adult strangers at all.
2. In one experiment, Paul Bloom and his fellow researchers presented 6-and-10-month-olds with a little morality play. The babies watched as a puppet would try to push a ball up a hill. Then, the babies saw one of two things happen. Either another puppet would come along and help the first puppet push the ball up the hill, or another puppet would show up and hinder the first puppet by pushing the ball down the hill.

After the babies watched these scenarios, the researchers presented each puppet to the babies. They wanted to see which puppet the babies would reach for. It turns out that nearly all of the babies, no matter how old they were, reached for the nice helping puppet. But are babies attracted to goodness or are they simply repelled by meanness? To find out, the researchers introduced a third character into the mix—a neutral one who neither helped nor hindered the main puppet. Then, they let the babies choose which puppet they wanted. The babies preferred the neutral character to the mean character, and the good character to the neutral character.

3. Even babies as young as three months showed moral awareness. They can control their eyes. You can tell what a baby likes by what it looks at. Researchers showed the three-month olds the same morality play with the helping and hindering puppets and then placed the puppets in front of them afterward. Most of the babies looked toward the nice puppet.
4. Beyond distinguishing between good and bad, young children also have an understanding of fairness and justice. In one version of the helping/hindering study, one of the babies actually reached over to the mean puppet and smacked it on the head.

The following findings are however also relevant:

5. From an early age, babies show bias to their in-group. Babies are quick to separate the social world into "us" versus "them." For example, if a baby is raised by a woman, it prefers to look at female faces; if it is raised by a man, it prefers looking at a male face; if it is raised by white parents, it prefers looking at white faces rather than black or asian ones.
6. The in-group bias manifests itself also in language. Minutes after they are born, babies prefer listening to people speaking the babies' mother tongue. Babies also prefer interacting with people who don't have uncommon accents.

It follows from the previous that humans are endowed with a moral sense (universal morality) that potentially extends to all humans (and even non-humans), but that they also have developed a moral sense that extends to a specific collectivity they identify with. This sense frequently appears to trump universal morality.

The question comes then up how altruistic behavior towards ingroups could have been evolving given that, according to neo-Darwinian views, genes and the behavior they influence are basically selfish. But what if humans are altruistic by nature? In that case, the question should be inverted and reformulated as follows: If animals and humans are altruistic by nature, why did selfishness, competitiveness and aggression evolve[4]?

Furthermore, if morality is innate, why is there such diversity in the social extension of morality: some societies demonstrate moral concern for all humans, others demonstrate it only for certain groups of humans, while some indigenous societies extend morality to animals, plants, mountains or the sea. Does this imply that Peter Singer's expanded circle has been more present in indigenous societies than it is in our society? Is Singer's circle expanding after all with the passing of time? If we look at indigenous societies that have been morally more inclusive than our current Western civilization, it appears even that Singer's circle is shrinking.

There are certain values that apparently are trans-geographically and trans-culturally accepted. They include a sense that helping someone in need is morally good, while harming innocent people is morally bad; that one should be good to those who have been good to her; that cheating is bad; an understanding that good people should be rewarded and bad people should be punished; that equal opportunities should be promoted. Moral emotions like empathy, guilt and righteous anger constitute part of those values.[5]

In the words of Paul Bloom in *Just Babies*: "There are two discoveries that I discuss in *Just Babies* that influence how I think about adult moral reasoning. The first is that there are hard-wired moral universals. To an important extent, all people have the same morality; the differences that we see—however important they are to our everyday lives—are variations on a theme. This universality… suggests that if we look hard enough, we can find common ground with any other neurologically normal human".[6]

And:

> The second discovery is the importance of reason. Prominent writers and intellectuals like David Brooks, Malcolm Gladwell, and Jonathan Haidt have championed the view that, as David Hume famously put it, we are slaves of the passions. Our moral judgments and moral actions are driven mostly by gut feelings—rational thought has little to do with it. I find this a grim view of human nature, but if it were true, we should buck up and learn to live with it.
>
> But I argue in Just Babies that it isn't true. It is refuted by everyday experience, by history, and by the science of developmental psychology. It turns out instead that the right theory of our moral lives has two parts. It starts with what we are born with, and this is surprisingly rich: babies are moral animals. But we are more than just babies. A critical part of our morality—so much of what makes us human—emerges over the course of human history

---

[4]https://www.psychologytoday.com/us/experts/darcia-f-narvaez-phd; last accessed on 1 November, 2020.
[5]https://www.scientificamerican.com/article/the-moral-life-of-babies/; last accessed on 1 November, 2020.
[6]Ibid.

and individual development. It is the product of our compassion, our imagination, and our magnificent capacity for reason.[7,8,9]

This brings us to the following question: *Apart from being innate, are moral values also universal?*

In the last couple of years in particular, a lot of people are thinking less about waistlines and paychecks and more about how the things they do matter in the wider world. A Marist Poll found "being a better person" was the most popular New Year's resolution for 2018. It was also the No. 1 resolution in 2017, marking a shift from the previous decade in which "losing weight" topped the list 80% of the time (in 2018 it tied).

"There's a crisis in the United States today, that too many of us have lost the sense of collective responsibility for our neighbors," said Rabbi Jill Jacobs, executive director of the nonprofit *T'ruah: The Rabbinic Call for Human Rights.*

Many people want to regain this collective responsibility. But what does this actually mean? What does it mean to be "good?" Social psychologists, ethicists and religious leaders say we see eye-to-eye on the big stuff. We believe it's good to be kind, fair and just; it's bad to cheat, murder and steal.

"The truth is that when you're talking broad strokes, no matter where you look, people value similar traits in character," says David Pizarro, a Cornell University professor who studies moral reasoning, judgment and emotion. According to him, evidence suggests we're all born with some innate sense of morality and fairness, which makes us sensitive to the distress of others.[10]

In the second part of this book I will cite dozens of thoughts written down in literature originating from very different cultures and from very different historical periods. All of them show a striking similarity with what we consider nowadays in the Western world as morally right.[11]

Why are there so many similarities among views on what is right and wrong in very different cultures? Is it so because morality is innate? We have seen that certain scholars argue precisely that. It can indeed be an explanation. The question is then how morality became innate.

There are various possible explanations of the emergence of morality: that our genes or God endowed us with it or that through evolution human morality evolved. This could have impacted on the moral sense we are being born with through epigenetic processes. In any case, there is no reason to limit the emergence and development

---

[7] Ibid.

[8] For Bloom's instructive video talk, see https://www.youtube.com/watch?v=MLrzetNHAYo; last accessed on 1 Noveber, 2020.

[9] For a useful addition to Bloom and the issue of morality of infants, see contributions referred to in https://www.psychologytoday.com/us/blog/am-i-right/201202/are-infants-moral; last accessed on 7 November, 2020.

[10] See https://www.usatoday.com/story/news/2017/12/26/you-good-person/967459001/; last accessed on 1 November, 2020.

[11] I owe gratitude to the Christian apologetic C. S. Lewis who assembled these thoughts in an addendum to his *Abolition of Man*. Although my arguments are very different from the ones presented by C. S. Lewis, his thoughts were a source of inspiration to me.

of morality to genes, epigenetics, God or evolution. A combination of some of these influences appears to be much more likely.

Reference

Bloom, Paul. 2013. *Just Babies: The Origins of Good and Evil*. New York, NY: Crown Publishers.

Chapter 2
Morality: Origin and Future

If evolution/natural selection is an explanation for the development of morality in humans, morality had been contributing to human evolutionary fitness. Close-knit ties between humans in small hunter-gatherer communities implied attitudes of solidarity within the community and a potentially hostile attitude toward other hunter-gatherer communities—"hunter-gatherer outgroups". The development of morality toward the ingroups was a consequence of the need to have a set of rules that was supposed to minimize harm in the community—harm that could primarily be inflicted by other hunter-gatherer communities. In that way morality contributed to safety and survival.

At a later stage, the need to make deals and have negotiations among competing hunter-gatherer communities (among else, to minimize harm they could inflict on each other) led to the extension of moral rules from one to other communities. Morality became more inclusive. This was the first phase of expansion of Peter Singer's circles of morality.

According to this evolutionary perspective, morality is not a social construct, but a product of natural selection and survival of the fittest. Those who had the most adaptive moral rules (e.g., the greatest good of the greatest number in the community they lived in) maximized their chances of survival, both as individuals and as a community.

The evolutionary perspective rejects the abolitionist ideas that morality is an illusion or fiction. If it were, it wouldn't have the role it is alleged to have in our evolution. On the other hand, one might ask whether morality is truly conducive to survival. It might be argued, namely, that morality hinders compromise. It may exacerbate conflicts rather than resolve them.

This line of reasoning is however incorrect in this context. Hindering compromise, namely, is not a morally apposite attitude *unless* the person(s) who hinder(s) compromise believe(s) that the compromise is immoral. Ideologues might hinder compromise because they believe that the compromise is against their ideology, but someone who hinders compromise because of its perceived moral failings, acts morally by rejecting the compromise. Another but related exception may be people with rigid characters (sometimes proponents of an ideology they identify with) who

© The Author(s), under exclusive license to Springer Nature Switzerland AG 2021
V. Rakić, *The Ultimate Enhancement of Morality*,
SpringerBriefs in Ethics, https://doi.org/10.1007/978-3-030-72473-3_2

cannot make adjustments to their strict moral rules, even if these adjustments are morally right. But in those cases, it is not morality that hinders compromise, but rigidity of character.

In fact, even if hindering an immoral compromise leads to a certain conflict, hindering it might still be morally right. Hindering such a compromise may be conducive to evolutionary survival of humans. Hence, there are strong reasons to argue that moral behavior played a rather significant role in the evolutionary survival of humans.

It is also possible to come up with the alternative that morality does not exist, apart from being an illusion, fiction or any other type of human construct. According to moral error theorists, all moral judgments are mistaken, as the world does not contain the properties and relations necessary for these judgments to be true. Hence, the correct stance to morality is moral abolitionism (Garner and Joyce 2018).

A variant of moral abolitionism is the notion that morality is not innate, that it has no genetic basis, but that it is a social construct. Humans adopt certain moral rules because of their embeddedness in a specific culture. Different cultures produce different moral values.

This interpretation, however, cannot account for the fact that certain moral values are shared irrespective of culture. Not to cheat on your friends, to help those in need, not to be bad to those who have always been good to you and various other values seem not to differ from culture to culture. They are trans-geographically and trans-historically valid. Those values that differ from culture to culture, on the other hand, are not true moral values. A more proper description of them is "conventions".

Although morality appears to be partially innate, it also played a role in the evolutionary development of humans. Being innate, it is part of human genetic make-up. But what was it that humans grounded their morality on through the ages?

Moral authority used to reside, and still resides in the views of some people, in one or more divine beings. In hunter-gatherer communities the power of the divinity was limited to these communities. With the development of agro-literate societies, it began to extend to larger populations. Ever increasing communication in the world between different cultures contributed to the development of monotheistic notions of an all-powerful God who was the supreme moral authority.

The Industrial Revolution and the emergence of democracy had as a consequence that states were not only in need anymore of select social groups (the aristocracy) but of as many citizens as possible. All citizens were potentially useful as workers, as voters and as soldiers. These social developments contributed to parts of the population in various countries gradually replacing the moral authority of the Holy Scripture with the moral authority of humanism.

In light of these developments in previous periods in human history, the question arises what the future of morality will look like. Will artificial intelligence and big data replace the moral authority of humans? Similar to humans replacing God, will AI and big data replace the moral relevance of humans? Will a human creation (AI, as the prime candidate) bring an end to humanity? If so, should we, and if we should, how should we prevent this?

The annihilation of the human species may have a moral justification if humanity is being replaced by morally superior beings. Humans being replaced by AI and big data has however no moral value. Much to the contrary, AI and big data, unlike humans, have no moral sense at this point of their development, and it is far from certain that they will ever acquire it. Humans being replaced by them would replace (the potential for) morality with something that is not morality.

The question is whether morality is something that ought to be preserved after all. Most humans have a moral sense. Goodness is valued very high among many humans. We don't know whether AI and big data are something humans should aspire to be replaced by. But we do know that moral rightness is by definition good. Hence, if we had to choose between, on the one hand, enhancement of humans leading to their replacement by AI and big data or by any other type of enhancement that is not moral enhancement, and, on the other hand, moral enhancement, we ought to opt for the latter. The reason is that the only type of enhancement that we know of as by definition morally right enhancement is moral enhancement. Consequently, the replacement of the human species by another sort of entity is desirable only if this entity is morally superior to humans. AI and big data are very far from that.

Hence, humanity is to aspire moral enhancement a priori, while the rightness of all other enhancements is debatable. In line with this, the replacement of humans with morally superior posthumans is morally desirable a priori, while their replacement with AI, big data or anything else is not necessarily morally desirable.

- Posthumans I define as a new species of beings who are morally superior to existing humans.

Existing human individuals don't have to be obliterated by morally superior posthumans. They are not being jeopardized as individuals. The human species being replaced by a morally superior posthuman species is therefore not detrimental to existing human individuals. It is detrimental only to the existing human species as a biological collectivity.

MBE might be a solution to enhancements that are detrimental to humans as individuals, as MBE could prevent humans inflicting serious harm upon themselves by unreasonable means of human enhancement. Moreover, it might be a solution to enhancements that are detrimental to humans as a species, if the existing human species is being replaced by a morally superior posthuman species. In both cases, human individuals are being preserved. In the first case, the existing human species is also being preserved. In the second case, the existing human species is being replaced by a morally superior species of posthumans. Both cases are morally justified.

- The difference between the existing human species and a posthuman species is that former is marked by a degree of moral conservatism, whereas the latter thinks and acts in accordance with a new morality—Ultimate Morality. Ultimate Morality will be discussed in more detail later.

- Cognition-related and behavior-related MBE has the potential to bridge the gap between the moral conservatism of existing humans and Ultimate Morality of posthumans.

Reference

Richard, Garner, and Richard Joyce. 2018. *The end of morality: Taking moral abolitionism seriously*. London: Routledge.

Chapter 3
Free Will and Moral Bioenhancement as a Compulsory Means of Avoiding "Ultimate Harm"

In this Section much of my attention will be devoted to the MBE theory advanced by Ingmar Persson and Julian Savulescu. The reason for that is that my own position can be better understood if contrasted to Persson and Savulescu. They are arguably the best known advocates of moral bioenhancement (MBE), but not for the reasons that have been addressed in the previous paragraphs. They ground their support for MBE on the argument that it will lower the likelihood of "ultimate harm".[1] Because of the severe consequences that ultimate harm entails, Persson and Savulescu favor making MBE compulsory.

Persson and Savulescu offer two crucial arguments for their stance. First, they suggest something that can be interpreted as an advocacy of the position that cognitive enhancement ought to be halted (or at least slowed down) until humans have become sufficiently morally enhanced. Second, they promote *compulsory* moral enhancement.

In Persson and Savulescu (2008) it is argued that moral enhancement ought to "accompany" other forms of enhancement, specifically *cognitive* enhancement:

> For if an increasing percentage of us acquires the power to destroy a large number of us, it is enough if very few of us are malevolent or vicious enough to use this power for all of us to run an unacceptable increase of the risk of death and disaster. To eliminate this risk, cognitive enhancement would have to be accompanied by a moral enhancement which extends to all of us, since such moral enhancement could reduce malevolence (Persson and Savulescu 2008: 166).

The argument that cognitive enhancement "would have to be accompanied" by moral enhancement, appears to imply that the former should be avoided until humans become sufficiently morally enhanced. In the words of Persson and Savulescu:

> Therefore, the progress of science is in one respect for the worse by making likelier the misuse of ever more effective weapons of mass destruction, and this badness is increased if scientific progress is speeded up by cognitive enhancement, until effective means of moral enhancement are found and applied (Ibid., 174; emphasis added).

---

[1] Persson and Savulescu define "ultimate harm" as an event or series of events that make worthwhile life on this planet forever impossible (Persson and Savulescu 2014, 251).

That cognitive enhancement ought to be *preceded* by moral enhancement might also follow from Persson and Savulescu's reference to C.S. Lewis's stories and the "Deplorable Word" (a magical curse which will end all life in the world except that of the one who pronounces it):

> If we all knew the Deplorable Word, the world would likely not last long. The Deplorable Word may arrive soon, in the form of nanotechnology or biotechnology. Perhaps the only solution is to engineer ourselves so we can never utter it, or never want to utter it" (Ibid., 175). In other words, humans ought to be morally "engineered" so that they will never be able to destroy themselves by the technological capacities they have.

Persson and Savulescu understand moral enhancement as our *motivation* to act morally (Ibid., 167). They offer the steady decrease in racism through human evolution as an example of such a motivationally determined understanding of moral enhancement: the role of racial distinction to signify a lack of kinship by marking off strangers from neighbors has been gradually losing its biological significance, enabling us to comprehend the moral falsity of racism (Ibid., 168). Since moral features are not a social construct, but are based in our biological makeup (Ibid., 168), Persson and Savulescu conclude that the potential hazards of cognitive enhancement are to be kept under control by a "vigorous research program on understanding the biological underpinnings of moral behavior". If these hazards can be controlled successfully, effective forms of moral enhancement are our duty and ought to be *mandatory* (Ibid., 174).

Persson and Savulescu (2008) ground the authors' MBE support on the argument that this type of enhancement will lower the risk of "ultimate harm". In Persson and Savulescu (2011) the argument of "ultimate harm" is elaborated in more detail. "Ultimate harm" can occur as a consequence of various factors, ranging from the use of weapons of mass destruction to catastrophic climatic changes. The underlying problem is that human moral psychology has been adapted to life in small, cohesive societies with primeval technology, while it is unprepared for the moral challenges of a technologically advanced global society. Life in traditional society has developed a bias towards the future among humans, disposing them to care primarily about immediately upcoming events that are relevant to them and their close neighbors. Furthermore, humans are still morally unprepared to respond appropriately to the hardships of larger groups. The development of advanced scientific technology appears to have resulted in the need for a radical change of human moral dispositions.

Hence, it essential that the possibilities of moral enhancement by means of genetic and other biomedical techniques be investigated. The misfit between a limited human moral nature and a technologically sophisticated global society ought to be ameliorated by moral enhancement, in order to achieve restraint, promote cooperation, develop respect for equality, as well as other values that are now necessary for the survival of humanity. And it is precisely scientific progress, the cause of this misfit, that might be employed to address it - by offering means leading to the enhancement of the morality of our behavior. But that is precisely where the caveat ("the bootstrapping problem") is: human beings, i.e. those who are in need of moral enhancement, are the ones who have to make a morally wise use of the techniques of moral enhancement (Ibid., 498).

That is how Persson and Savulescu arrive again at the conception of making MBE compulsory. In Persson and Savulescu (2008) they advocated compulsory MBE openly: if hazards with the potential of causing "ultimate harm" can be controlled successfully, "effective forms of moral enhancement are our duty and ought to be mandatory" (Persson and Savulesu 2008, 174). In their later writings (e.g., Persson and Savulescu 2012), they didn't insist anymore on making MBE mandatory, although from much of what they argue compulsory MBE is being implied. For example, the implication of the above mentioned "bootstrapping problem" is either to abort MBE, to continue to advocate compulsory MBE, or to give arguments in favor of VMBE that might circumvent the "bootstrapping problem". As Persson and Savulescu have not aborted the idea of MBE, nor have they given any reasons favoring VMBE, we can conclude from this that they still are in support of making MBE mandatory.

Persson and Savulescu (2008) can also be criticized from the following perspective—one taken by John Harris and Elisabeth Fenton. If moral enhancement is to take place at a biological level, non-traditional cognitive enhancement is required. Consequently, if we do not continue scientific research into enhancement, we have no hope of achieving the great moral progress that will ensure humans lowering the likelihood of ultimate harm. However, the argument goes, the logic in Persson and Savulescu (2008) appears to lead to an "obstinate predicament": "Scientific progress is both the means of our salvation, as well as the means of our downfall" (Fenton 2011: 148).

In line with Persson and Savulescu's notion of compulsory MBE, the "god machine" is imagined as a mechanism that is designed to *impose* morally laudable behavior. Hence, it is entirely in line with a program of compulsory MBE. It is left to every individual to decide for herself whether she wishes to be connected to this device. In that regard, it might appear to be respectful of our free will. But such an impression deceives us. Unlike medication for MBE that we decide to take and can equally decide stop taking (unless we get addicted to it), the "god machine" hijacks our free will (or that what we believe is our free will) once we get connected to it. This device is charged with policing our thoughts in order to keep us away from acting immorally. Unlike God from the Judeo-Christian and Islamic traditions who keeps our free will intact, the "god machine" resembles more a "police machine" than anything we associate with God from those traditions.

There is also no doubt that the "god machine" would be disinclined to accept our decision to disconnect ourselves from it. Hence, the outcomes of our voluntary decision to take MBE medication and our voluntary decision to connect to the "god machine" are very different. In the first case, our free will remains intact (unless, again, we become addicted to the MBE drug we have been administered), while in the second case our free will is being lost.

Making MBE compulsory in order to lower the likelihood of ultimate harm would deprive humans of their freedom of the will. Depriving humans of this freedom means taking away something that is essential for the existence of humans as moral beings. In actual fact, compulsory MBE, albeit intended to avoid "ultimate harm", already inflicts a degree of ultimate harm on humans by depriving them of an essential human quality. As the "god machine" is an instrument of compulsory MBE, it is a device designed (unintentionally) to inflict harm on humans—if not ultimate, certainly major harm![2]

Persson and Savulescu replied to my critique of compulsory MBE (Rakić 2014) by arguing that freedom is a matter of degree (Persson and Savulescu 2014). Michael Selgelid (2014) found my concept alien to "scalar bioethics". I agree that there are degrees of freedom if we understand freedom as a political concept. We can have more or less free elections, more or less free media. Freedom of the will, however, is a threshold concept. Once limitations are imposed on what we are allowed to will, we cannot call our will free anymore. As soon as an external mechanism decides what we are permitted to will, our freedom of the will has not been limited "to a degree". Our will has ceased to be free.

It is of course possible to argue that a free will does not exist, that it is an illusion. The Libet experiment, as well as later experiments with similar findings, suggest that our decisions might take place before we become aware of them (Libet at al. 1983, 1986). More recent findings presented by Lau and colleagues suggest that the perception of intention occurs after executive motor movements (Lau et al. 2007). Wegner reasons along similar lines when discussing how auditory hallucinations produced by schizophrenia seem to suggest a divergence of will and behaviour (Wegner 2003). Kühn and Brass argue that we might even be unable to veto or halt a decision we have made unconsciously, as this veto might also have taken place on an unconscious or subconscious level (Kühn and Brass 2009). It should be noted, however, that those are internal limitations to our free will. Furthermore, they are limitations that we are not aware of in our direct experience. Compulsory MBE, on the other hand, involves an external mechanism that is designed to affect our will, a mechanism we are conscious of.

The implication is that compulsory MBE, an enterprise that affects what we perceive as our free will, also affects what we perceive as our human identity (which includes us having a free will). Hence, compulsory MBE, affecting what we perceive as our free will, runs contrary to our notion of who we are. *In that sense, it inflicts another essential harm as well: harm, possibly ultimate harm on our identity as human beings.*

It is of course possible to redefine this identity. However, if we do so by denying the reality of our experience of free will, the redefinition of our human identity, even if possible, would be both difficult and highly traumatic. Hence, it is something that is to be avoided.

---

[2] A less ambitious and presumably less beneficial/detrimental technique than the "god machine" is neuromarketing. Nevertheless, it can acquire some dispositions of the "god machine". Neuromatketing uses clinical information about cerebral processes in order to understand what can be expected when consumer behavior is concerned. Experts in neuroimaging analyze brain responses to different stimuli and predict how consumers will react to them. Neuromarketing can also be applied in ethics. Certain types of moral behavior can in that way be stimulated or inhibited. "***Neuromarketing of morality***" might in that sense be a precursor to the "god machine". For the possibly inherent evil of marketing, neuromarketing, marketing directed to vulnerable groups, see https://en.wikipedia.org/wiki/Marketing_ethics (last accessed on 7 November, 2020).

Moreover, there is evidence suggesting that the belief that we have in a free will is a significant motivation for us to act morally. Various empirical findings substantiate this. Baumeister and colleagues point to findings that a disbelief in free will decreases helpfulness and increases aggression (Baumeister et al. 2011). Elsewhere, Baumeister et al. argue that trust in free will has behavioral consequences, including increases in socially and culturally desirable acts (Baumeister et al. 2009). In one publication Rigoni et al. show that the readiness potential for acting is lowered in individuals induced to be skeptical about a free will (Rigoni et al. 2012a). In another article Rigoni and colleagues demonstrate that undermining free will can degrade self-control and that it leads to other antisocial tendencies (Rigoni et al. 2012b). Vohs and Schooler provide evidence that mistrust in free will increases the tendency to cheat (Vohs and Schooler 2008). All these findings further strengthen the argument that even the illusion that our will is free should not be easily abandoned. If we abandon it, we might be less prone to act morally. If we believe that freedom of the will is a matter of degree, that we do not fully possess what we have always experienced as a free will, we will be less likely to even try to act morally, achieving exactly the opposite of our goal of moral enhancement.[3]

Furthermore, as the "god machine" deletes our *thoughts* that it considers as "grossly immoral", it not only limits our freedom of the will, but also our freedom of thought. It is therefore no wonder that in *Unfit for the Future* (2012) Persson and Savulescu put forward their reservations toward liberal democracy. Indeed, compulsory moral bioenhancement requires a degree of authoritarianism. The "god machine" cannot function in a liberal social setting.

It is not clear who is to decide what qualifies as a "grossly immoral" thought. Let us assume that it is the "moral elite" in a society. Or just a few people who know best where the line between immoral and "grossly immoral" is to be drawn. How can we know who they are? What is the moral elite? Moreover, why should we believe that this elite or just a few of the morally most proper people (whoever they possibly can be) would have the power to be in charge of the "god machine"? Or, for that matter, to be in charge of any type of compulsory MBE program? It is far from certain that the moral and political & financial elite will be congruent. They are likely to be different people. In that sense, the "god machine" cannot be brought into practice. If it were ever developed, it would be a device under the control of the most powerful groups in a society. The same holds for any other compulsory MBE program: it would be run by the most influential social groups, which are by no means necessarily the "most moral" social groups.

Last but not least, compulsory MBE brings into question the conception of love (as advocated in Rakić 2021). If love is a matter of our will, as has been argued in Rakić (2021), compulsory MBE, infringing on our will, would also infringe on our (capacity to) love. If a "god machine" decides which types of love are acceptable and which are "immoral", it decides who deserves our love and who not. In that case it is not we who love. The "god machine" loves instead of us. It loves in our name.

---

[3] The mentioned findings have also been critically assessed by some authors. A discussion of these critiques is beyond this chapter's scope.

Freedom and love are essential components of morality. As compulsory MBE diminishes both our freedom and our full capacity to love, also bringing into question our human identity and moral reflection, it is highly detrimental to morality. It achieves exactly the opposite of moral enhancement: moral decline. That is, in short, my arguments against the position of Persson and Savulescu.

References

Agar, Nick. 2004. *Liberal Eugenics: In Defence of Human Enhancement*. Hoboken, NJ: Wiley-Blackwell.
Agar, Nick. 2013. Why is it possible to enhance moral status and why doing so is wrong? *Journal of Medical Ethics* 39: 67–74.
Agar, Nick. 2014. Truly human enhancement: a philosophical defense od limits. *Theoretical Medicine and Bioethics*: 1–4.
Agar, Nick. 2015a. *The Sceptical Optimist: Why Technology Isn't the Answer to Everything*. Oxford: Oxford University Press.
Agar, Nick. 2015b. Moral bioenhancement and the utilitarian catastrophe. *Cambridge Quarterly of Healthcare Ethics* 24 (1): 37–47.
Agar, Nick, and J. McDonald. 2017. Human enhancement and the Story of Job. *Cambridge Quarterly of Healthcare Ethics* 26 (3): 449–458.
Baumeister, R.F., E.J. Masicampo, and C.N. DeWall. 2009. Prosocial benefits of feeling free: Disbelief in free will increases aggression and reduces helpfulness. *Personality and Social Psychology Bulletin* 35: 260–68.
Baumeister, R.F., A.W. Crescioni, and J.L. Alquist. 2011. Free will as advanced action control for human social life and culture. *Neuroethics* 4: 1–11.
DeGrazia, David. 2014. Moral enhancement, freedom, and what we (should) value in moral behaviour. *Journal of Medical Ethics* 40 (6): 361–368.
Douglas, Thomas. 2008. Moral enhancement. *Journal of Applied Philosophy* 25 (3): 228–245.
Fenton, Elizabeth. 2011. The perils of failing to enhance: A response to Persson and Savulescu. *Journal of Medical Ethics* 36: 148–151.
Harris, John. 2011. Moral enhancement and freedom. *Bioethics* 25 (2): 102–111.
Kühn, S., and M. Brass. 2009. Retrospective construction of the judgement of free choice. *Consciousness and Cognition* 18: 12–21.
Lau, H.C., R.D. Rogers, and R.E. Passingham. 2007. Manipulating the experienced onset of intention after action execution. *Journal of Cognitive Neuroscience* 19: 81–90.
Libet, Benjamin, C.A. Gleason, E.W. Wright, and D.K. Pearl. 1983. Time of conscious intention to act in relation to onset of cerebral activity (readiness-potential). The unconscious initiation of a freely voluntary act. *Brain* 106: 623–42.
Libet, Benjamin. 1986. Unconscious cerebral initiative and the role of conscious will in voluntary action. *Behavioral and Brain Sciences* 8: 529–66.
Persson, Ingmar, and Julian Savulescu. 2008. The perils of cognitive enhancement and the urgent imperative to enhance the moral character of humanity. *Journal of Applied Philosophy* 25 (3): 162–177.
Persson, Ingmar, and Julian Savulescu. 2011. Unfit for the future? Human nature, scientific progress, and the need for moral enhancement. In *Enhancing Human Capacities*, edited by Julian Savulescu, Ruud Ter Meulen and Guy Kahane (pp 486–500). Oxford: Wiley-Blackwell.
Persson, Ingmar, and Julian Savulescu. 2012. *Unfit for the Future*. Oxford: Oxford University Press.
Persson, Ingmar, and Julian Savulescu. 2014. Should moral bioenhancement be compulsory? Reply to Vojin Rakic. *Journal of Medical Ethics* 40 (4): 251–252.

References

Rakić, V. 2014. Voluntary moral enhancement and the survival-at-any-cost bias. *Journal of Medical Ethics* 40 (4): 246–250.

Rakić, V., and H. Wiseman. 2018. Different games of moral bioenhancement. *Bioethics* 32 (2): 103–110.

Rakić, V. 2021. *How to Enhance Morality*. Dordrecht: Springer.

Rigoni, D., S. Kühn, G. Sartori, and M. Brass. 2012a. Inducing disbelief in free will alters brain correlates of preconscious motor preparation: The brain minds whether we believe in free will or not. *Psychological Science* 22: 613–18.

Rigoni, D., S. Kühn, G. Gaudino, G. Sartori, and M. Brass. 2012b. Reducing self-control by weakening belief in free will. *Consciousness and Cognition* 21: 1482–1490.

Selgelid, Michael J. 2014. Freedom and moral enhancement. *Journal of Medical Ethics* 40 (4): 215–16.

Sparrow, Robert. 2014. Egalitarianism and moral enhancement. *American Journal of Bioethics* 14 (4): 20–8.

Verkiel, Saskia. 2017. Amoral enhancement. *Journal of Medical Ethics* 43 (1): 52–55.

Vohs, K.D., and J.W. Schooler. 2008. The value of believing in free will: Encouraging a belief in determinism increases cheating. *Psychological Science* 19: 49–54.

Wegner, D.M. 2003. The mind's best trick: How we experience conscious will. *Trends in Cognitive Sciences* 7: 65–69.

Wiseman, Harris. 2016. *The Myth of the Moral Brain*. Cambridge, MA: MIT Press.

Chapter 4
Past and Present of Happiness and Morality

In Rakić (2018) I have argued that the grounding rationale for moral enhancement ought not to be the lowering of the likelihood of "ultimate harm", as Persson and Savulescu assert, but rather happiness. Akin et al. (2009) showed that morality and happiness operate in a circularly supportive fashion (in a positive feedback loop) for most people most of the time: the more moral they are, the happier they will feel; the happier they feel, the more moral they will behave. The correlation between happiness and morality has also been shown in other studies, including Dunn et al. (2008), Isen and Levin (1972) and Sheldon and Lyubomirsky (2004). In what follows, some essential approaches to happiness through the ages will be briefly reviewed. Afterwards, the focus will again be on the relationship between happiness and morality.

a. *A Brief History of Happiness*

A lot of people think about happiness as about something that will emerge in the future: "I will become rich, I will lose weight, I will be better to other people, I will work less, I will visit the Seychelles….Jasper Bergink, on the other hand, emphasizes that the future is a highly unreliable source of happiness. It is unreliable because it is so unpredictable. He argues that the source of happiness is the past. If we are able to revive those events that made us happiest during our lives, we make them part of the present—as happy experiences and as sources of inspiration for happiness in the present.[1]

We will take a brief look now at how happiness has been perceived in a number of philosophical worldviews through the ages.

Confucianism and Taoism

Confucius was a moralist, a preacher, an educator. Unlike Confucius, the Taoists insisted that inner harmony will achieve more than any preaching of morality. Confucius was an advocate of moral education, whereas Taoists focused more on the

---

[1] https://www.forastateofhappiness.com/happiness-in-the-past-present-and-future/; last accessed on 7 Novemner, 2020.

motivational aspect of being good. This motivation resides in virtue, while virtue leads to happiness. This ancient approach already introduces the idea of virtue being conducive to happiness.

Confucian morality respects noble conventions, devaluating power and financial success. The moral human will behave virtuously, irrespective of his power and richness. Hence, already Confucianism insisted on morality and power being in a tense relationship.

Confucius thought of fellow people as people with similar needs. These needs one should respect as much as one's own. One should treat fellow people in a similar way as she treats herself. This is a position that can be related to the Golden Rule—one of various examples of similar moral standards being present trans-geographically and trans-historically.

According to Confucian standards, the virtuous person does not only devaluate power and richness, but he is also not guided by sensual pleasure. The pleasure by which the virtuous person is guided is the pleasure in exercising virtue. In order to achieve this, education and self-examination are essential (Annping Chin 2008).

Taoist thought is not directed to the elite (as Confucianism is), but to the common human, irrespective of her position. For Taoists, the Tao itself is beyond everything. In fact, it is "nothing". But this "nothing" is tremendously powerful, as everything that exists derives its existence from it. Hence, humans cannot be guided by virtue or morality, as all their concepts suppose the knowledge of its opposite: "Every concept of beauty in the world is connected to ugliness. Every notion of what is good supposes the knowledge of evil" (Lao Zi 2010, 2). Consequently, the Tao is the ultimate moral guideline.

Morality is not meaningless, but is only secondary to the Tao. If all humans were to follow the Tao, morality would not even be necessary. Following Tao means following the natural course of things. But this does not mean following our desires. The Tao is beyond desires. It consists of "acting by not-acting", a conception known as *Wu Wei*.

Wu Wei is largely based on acting without the desire to obtain personal benefit from it: "Tao exists in never doing anything. If a ruler were to contain the Tao, all beings would spontaneously become civilized. And if they were civilized and wanted to start acting, I would make them peaceful by the simplicity of the nameless, of the nameless unprocessed wood. Yes, I would take care that they were without desire. Because without desire, and through silence, the world will spontaneously come to rest" (Lao Zi 2010, 37).

Taoism does not advocate however to be *in everyday life* without any desires *at all*. It is excessive desires that are not virtuous. In certain aspects similar to Ancient Greeks, Taoists advocate moderation: "That is why the Wise avoids all exaggeration, profusion and presumptuousness" (Lao Zi 2010, 29). And: "They who hold this Tao do not attempt to achieve their limits" (Lao Zi 2010, 15). In the case of war: "It is best to succeed and leave it at that. Don't try to profit from the situation to get more by violence" (Lao Zi 2010, 30); "Even if he wins the war, he will not hold a parade, but a funeral service" (Lao Zi 2010, 31).

Although neither Confucians nor the Taoists advocate lives directed to the fulfillment of desires, they do give desires a role as means to (different) ends. Confucians attempt to direct desires at virtuous thoughts and behavior, while Taoists focus more directly on happiness and the state of mind leading to it. All in all, it can be concluded that ancient Chinese philosophers have already seriously addressed the relationship between happiness and morality.

Greeks and Romans

Aristotle posited that all humans strive for happiness (in the sense of *eudaimonia*[2]), but differ on the issue what is to be understood as happiness/*eudaimonia*. Atistotle considers happiness as the ultimate human objective, at which humans aim only for itself. Hence, happiness is *not* a means to something else.

An understanding that can be found in ancient Greek thought is that happiness is durable, within us, independent of external circumstances. Power, wealth, reputation/status or even fame are not lasting sources of happiness, because they depend, in the final instance, on means that are outside our control (Aristotle 1998).

For Aristotle, happiness and virtue of character (*ethike arete*) are intertwined. Namely, happiness can be achieved by virtuous actions. But virtue must be practised. Once learned, our desires will experience a transformative change: we will experience pleasure by acting virtuously. Hence, by acting morally humans can obtain gratifying experiences. Happiness is an essential gratifying experience. After a process of education, virtuous actions will therefore enhance human happiness and humans will understand that they are in their self-interest (Aristotle 1998).

Although these thoughts resemble to bear similarities with Confucianism and Taoism, Aristotle also emphasizes the importance of practical matters for happiness. Similar to a flute player who cannot exercise his virtue without a flute, any human needs friends and life in a polis to exercise essential human virtues. Hence, human happiness depends to some degree on certain practical matters—a thought that was not present in that form among the ancient Chinese (Aristotle 1998).

All this brings Aristotle rather close to contemporary happiness research. Still, in Aristotle's understanding of happiness, "feelings" do not figure most prominently. Instead of our utilitarianism-inspired weighing of pleasures and pains (the "felicific calculus"), Aristotle considers virtue and happiness in the first place to be a proclivity of some persons, as well as their activities based on this proclivity, rather than a feeling. In contrast, modern happiness researchers largely value happiness because of the feelings it can bring about.

The Stoics do not depart from Aristotle (and the ancient Chinese) in that they see a close connection between happiness and virtue. The Stoics seem however in one respect to be closer to the ancient Chinese than to Aristotle: they regard happiness as something that is largely independent of circumstances. Moreover, Aristotle regards happiness as an activity, while the Stoics understand it as a mental state. This mental

---

[2]*Eudaimonia* is an Ancient Greek term denoting a stable type of happiness. It is also commonly translated as welfare, blessedness, human flourishing and prosperity.

state they call *ataraxia*. Ataraxia characterizes tranquillity—as freedom from any disturbances.[3]

The Stoic argument runs as follows. When we satisfy a desire, we derive pleasure from it. But there is no certainty that our desires will be satisfied. An unsatisfied desire causes pain (psychological pain, sometimes even physical pain). Having no guarantee that we will experience pleasure rather than pain, and fearing being deprived from that what presently brings us pleasure, we cannot be happy. Hence, we should aspire happiness that is stable. Such a stable state of happiness can be achieved only by giving up desire. Without desire, we will be deprived of pleasures, but also of pains. In return, our mind will be free from disturbances (Brennan 2005).

As a matter of fact, both in ancient Chinese philosophy, as well as in ancient Greek and Roman philosophy, there is a close relationship between morality and happiness. Moreover, all mentioned views advocate that humans ought to limit and master their desires. These desires are either to be overcome (Taoists and Stoics), or to be directed to moral actions (Confucius and Aristotle). The former approach insists on humans trying to find happiness and virtue within themselves, while the latter approach is more inclined to emphasize the relevance of practice, being true to the conception of humans remaining dependent on the outside world.

Modernity

Unlike the ancient traditions (Chinese, Greeks, Romans) that focused on happiness as *moderation of desire*, modern Western philosophy considers happiness primarily as *gratification of desire*. Furthermore, happiness as the gratification of desire became in modern Western philosophy the groundwork of morality.

This change was not visible yet in Spinoza and Leibniz, for whom virtue and happiness were still strongly related to each other. Desires had to be managed in such a way that they lead to a life that is both virtuous and happy. But with Grotius and Hobbes this approach changed into the question how people who are guided by desires and passions can live together in peace.

Desire became the platform on which morality rests.

Hobbes' desires have the following features:

- They fulfill the role of the "forces" Newton had discovered in physics. Similar to the forces in nature, desires are causal forces that operate at a human level.
- Since humans are equal, their desires are equally valuable. The ancient distinction between desires of lower and desires of higher order disappears.
- Happiness is being reduced to the fulfillment of desire (Hobbes 2008).

The third feature opens the space for the utilitarian extension of this position: moral is what makes most people happy to the highest extent, that is, what fulfills their desires. Ancient elitism, with *moderation of desires* being essential to virtue

---

[3] Ataraxia is also aspired by Epicureans. The Epicureans challenged however the conception of virtue and happiness being almost one and the same, in that they considered virtue as a means leading to happiness. In that respect they came quite close to modern thoughts on happiness. I will return to this later.

and happiness, disappears. Happiness and morality become tightly connected to the *fulfillment of desires*. They become a matter of the masses. Morality is not reserved anymore for the few, but belongs to all.

The conception of morality as the fulfillment of desires has however its weaknesses. For example, self-sacrifice for moral reasons goes against our desires, it causes pain, and is therefore a threat to happiness and hence to morality. Suffering and disappointments from which we can learn, or giving meaning to losses, have no place is such a conception of morality, because they jeopardize the fulfillment of our desires and cause pain. Hence, they are bad—bad for our happiness, bad in moral terms. They are to be avoided.

Kant departs from this tradition, asking the question how people driven by desires can still live in peace with each other, and most importantly, *answering* this question by invoking the conception of respect to the law people give to themselves. Differently from the British tradition, in which desires determine the requirements of morality, Kant envisions a morality that goes beyond the objectives of our desires.

In Kant, the ties between happiness and morality are being cut. Following their natural inclinations, says Kant, humans aspire happiness through the fulfillment of their desires. Their practical reason steps however in there, in order to correct them and oblige them to act morally. Importantly, these moral actions are not related to human happiness (Beiser 2006).

It is also important to note that in Modernity morality has become an inter-subjective matter. Whereas ancient conceptions of a virtuous and happy life were more inward-oriented, Modernity treats morality as something that should be valued on the basis of its inter-subjective merits. The ancient ethical question of how to deal with myself has been transformed into the modern ethical question how to deal with others. Modernity has moved ethics from the subjective to the inter-subjective sphere.

The relationship between happiness and morality reminds us therefore also of the question whether morality is necessarily inter-subjective. If we detach morality and happiness from each other, morality extends to the inter-subjective sphere only. But if we return to the ancient notion of this relationship, morality resides in the sphere of subjectivity as well. For example, the ancient conception of moderation of desire (key to happiness) refers to us being moral toward ourselves. The modern conception of gratification of desire, on the other hand, opens up the sphere of being moral to others.

This issue is tightly connected to the question of what happiness entails. Is happiness gratified desire? The modern (largely British) tradition would principally concur with this. Hobbes is a paramount example: desires are rooted in passions, humans aspire to achieve the objects of their passions and once they achieve them, they are happy. This type of happiness is frequently called "hedonic happiness" (how happy we are at the moment). Conversely, the ancient understanding of happiness as moderation of desire refers to a different type of happiness. Happiness is *eudaimonia*: a stable feeling. In contemporary terminology it would be "overall life satisfaction".[4] If we regard desires as passions, they frequently are an impediment to *eudaimonia*. Then they are detrimental to our peace of mind.

In conclusion, the difference between the ancient notion of happiness as moderation of desire and the modern notion of happiness as gratification of desire, reveals two different approaches both to the understanding of happiness and to the understanding of morality. In the ancient notion happiness is largely inside ourselves, so that morality (virtue) is something that frequently refers to us, to what kind of moral characters we are. In the modern notion of happiness as the gratification of desire, happiness is largely outside ourselves. Consequently, morality is fundamentally inter-subjective.

Happiness Now

Contemporary happiness research shows a tendency to reconcile happiness inside ourselves (internally caused happiness) with happiness that is caused by factors that are outside ourslevs (externally caused happiness).

Sheldon and Lyubomirsky (2004) offers findings about the objectives people can choose. Happiness is enhanced when people opt for objectives that are:

- feasible, realistic and attainable;
- those they are making progress towards;
- personally meaningful;
- those that people feel highly committed to;
- intrinsic;
- concerned with community, intimacy and growth;
- valued by one's culture;
- not conflicting;
- self-concordant and congruent with people's motives and needs.

These types of happiness are largely independent from external factors. They are independent types of happiness. Independent happiness resembles Stoic *ataraxia*, Taoist *wuwei*, and what Buddhists called *enlightenment*. It is important to note that contemporary happiness research tends to return to these ancient notions of independent happiness.

It might indeed often be in our best interest to turn to ancient notions of happiness/morality.[5] This can entail:

---

[4]It deserves notice that "overall life satisfaction" is not to be confused with *eudaimonia*. They are not the same types of happiness. Both are related however in that they do not refer to immediate pleasures, but to long-term satisfactions.

[5]A number of Modernists did not follow the dominant thought of their time, in which happiness was treated as the fulfillment of desires. A succinct formulation of such an approach can also be found in Thomas Jefferson's statement: "Happiness [is] the aim of life. Virtue [is] the foundation of happiness" (https://www.psychologytoday.com/us/blog/am-i-right/201901/why-are-times-so-hard-today; last accessed on 1 November, 2020).

(1) to desire less (i.e. to have a smaller number of desires);
(2) to desire what we desire with less intensity;
(3) to desire things that are less dependent on outside resources;
(4) to desire other people's self-interest.[6]

This view differs substantially from the modern view of happiness, which became the conventional view, according to which happiness is largely the gratification of desires. Conflicting desires for scarce resources are frequently at the root of conflicts. In political science and economy, they are also the point of departure for much of moral thinking, starting to a large part with Hobbes.

What follows is a number of brief references to contemporary experiments that show morality and happiness in a mutually supportive relationship.

b. *Contemporary experiments on happiness*[7]

Kennon Sheldon and Sonja Lyubomirsky argue that we can become happier if we decide so. There are certain volitional or activity changes that can increase our happiness. They include resolving to regularly count one's blessings, pursue meaningful personal goals, or commit random acts of kindness (Sheldon and Lyubomirsky 2004). Elsewhere, Lyubomirski adds to this list of happiness stimulating and happiness sustaining activities the following: making someone else happier, affirming significant values, visualizing a positive future, and savoring positive experiences—in order to durably increase a person's happiness level beyond his or her "set point".[8]

Lyubomirsky endeavours to develop a "science of human happiness". To this end, she focuses on three key questions:

(1) What makes people happy?
(2) Is happiness a good thing?
(3) How can we make humans happier than they are?

Dunn et al. (2008) argue that spending money on others promotes happiness: "Although personal spending is of necessity likely to exceed prosocial spending for most North Americans, our findings suggest that very minor alterations in spending allocations—as little as $5 in our final study—may be sufficient to produce nontrivial gains in happiness on a given day" (Dunn et al. 2008: 1688). The authors also make an attempt to explain why people don't introduce corresponding changes in their behavior, as well as how to help them in that regard:

> Why, then, don't people make these small changes? When we provided descriptions of the four experimental conditions from our final study to a new set of students at the same university (N = 109) and asked them to select the condition that would make them happiest, Fisher's Exact Tests revealed that participants were doubly wrong about the impact of money on happiness; we found that a significant majority thought that personal spending (n = 69) would make them happier than prosocial spending (n = 40) (P < 0.01) and that $20 (n =

---

[6]This is not far from Sheldon and Lyubomirski (2004) which is being discussed in the section that follows.

[7]See Rakić (2018).

[8]See https://themythsofhappiness.org/about-the-author/; last accessed on 1 November, 2020.

94) would make them happier than $5 (n = 15) (P < 0.0005). Given that people appear to overlook the benefits of prosocial spending, policy interventions that promote prosocial spending— encouraging people to invest income in others rather than in themselves—may be worthwhile in the service of translating increased national happiness (Dunn et al. 2008: 1688).

Isen and Levine (1972) provide evidence for this: when we feel good we are more inclined to help others. In two studies experimenters induced subjects to feel good by offering them minor pleasures: one group received cookies while studying in a library, whereas members of the other group "incidentally" found a dime in the coin return of a public telephone. It turned out that members of both groups were more helpful than control subjects (Isen and Levine 1972: 384). It also turned out to be the case that subjects became more helpful not only when their good mood was brought about by another person (who handed out cookies), but also when their mood was enhanced in an impersonal manner and by a seemingly accidental event (finding a dime) (Ibid., 386, 387).

Hence, happiness is a good thing for two reasons:

1. Happiness is intrinsically good because it feels good to be happy.
2. Happiness tends to stimulate goodness, that is, morally appropriate behavior.

Related to reason 2: conversely, goodness tends to stimulate happiness. That brings us to an answer to Lyubomirsky's first question, that is, the question what it is that makes people happy. According to the findings that were discussed in the foregoing paragraphs, it is goodness that tends to make people happy. In line with this, we arrive at an answer to Lyubomirsky's third question, that is, how people can be made happier than they are. This can be achieved by making humans better, by enhancing their morality. In other words, moral enhancement can make humans happier still.

Conversely, as has been shown in Anik et al. (2009) and in Isen and Levine (1972), happiness tends to contribute to moral behavior.

Hence, the discussed contemporary findings about the relationship between morality and happiness, brings us back to Ancient notions dealing with this relationship. It links these findings not only to Aristotle and the Stoics, but also to Confucius and the Taoists. In their conceptions, happiness and morality were also intertwined in various ways.

References

Anik L, Aknin LB, Norton MI, Dunn EW. 2009. *Feeling good about giving: The benefits (and costs) of self-interested charitable behavior*. Harvard Business School Working Paper.
Annping C. 2008. *Confucius: A life of thought and politics*. New Haven and London: Yale University Press.
Aristotle. 1998. *The nicomachean ethics*. Oxford: Oxford University Press.
Beiser, F.C. 2006. Moral faith and the highest good. In *The Cambridge companion to kant and modern philosophy*, ed. Guyer Paul. Cambridge: Cambridge University Press.
Brennan, T. 2005. *The stoic life. emotions, duties & fate*. Oxford: Clarendon Press.

References

Dunn, E.W., L.B. Aknin, and M.I. Norton. 2008. Spending money on others promotes happiness. *Science* 319: 1687–1688.
Hobbes, T. 2008. *Leviathan*. New York, NY: Simon & Schuster.
Isen, A.M., and P.F. Levin. 1972. Effect of feeling good on helping: Cookies and kindness. *Journal of Personality and Social Psychology* 21 (3): 384–388.
Lao, Zi. 2010. *Het boek van de Tao*. Amsterdam: Uitgeverij Augustus. (All quotes translated into English by Kristofer Schipper).
Sheldon, K.M., and S. Lyubomirsky. 2004. Achieving sustainable new happiness: Prospects, practices, and prescriptions. In *Positive psychology in practice*, ed. A. Linley and S. Joseph, 127–145. Hoboken, NJ: Wiley.

Chapter 5
The Future of Happiness and Morality, Psychopathy and Utilitarianism

5.1 The Future of Happiness and Morality

Harari (2015) foresees a future marked by technological developments, without morality being a necessary corrector to these developments. This is exactly the opposite of the view of Persson and Savulescu who insist on MBE, even compulsory MBE, in order to prevent these technological developments to result in humanity experiencing "ultimate harm". Moreover, Harari envisions AI and big data gradually replacing humanity, without offering a view how this may be prevented. My prevention proposal is voluntary MBE, with happiness being its grounding rationale. This proposal builds on the positive correlation and circularly supportive relationship between happiness and morality, which has been discussed in the previous chapter: for most humans it turns out to be the case that the more moral they are, the happier they will be, *and*, the happier they are, the more moral will they be.

MBE can therefore moralize the future. This possibility is not taken into account by Harari. In opposition to him, I argue that the eventuation of a new superior species will not necessarily consist of merely technologically or cognitively enhanced humans (most likely enriched by their possession of AI and/or big data), but of posthumans who are also morally enhanced and happier.

The future can take two possible paths: 1. humans can be replaced by immoral machines/organisms (organisms, AI, big data or various (other) types of robots); their happiness levels can be technologically induced (in the case of organisms also brain stimulated) OR 2. humans can be replaced by morally enhanced posthumans with a capability of experiencing deeper, more complex and more differentiated types of happiness (ancient types of happiness notions should be included here, as has been shown in the previous chapter).

As has been indicated already, the correlation between morality and happiness might possibly apply to different types of happiness. One type is a state of subjective wellbeing (SWB) that is rooted in life satisfaction, while the other is based on positive affect. The former refers to stable states, the latter to more transient states.

In Correia et al. (2009) it is has been demonstrated that Belief in just world (BJW) and SWB are positively correlated, but only in case of SWB as life satisfaction and not as positive affect. In the words of the authors about their experiments:

> The results showed that there were no effects of experimentally induced mood changes on BJW (Study 1), but the saliency of life satisfaction reinforced BJW (Study 2). There was also an increase in life satisfaction when BJW was activated (Study 3) but the activation of BJW had no impact on the mood of the participants (Study 3). In our view, these results stress two important points... (a) they show that transient affective states do not influence or are influenced by BJW, and that this relationship is present only for the more stable affective dispositions; and (b) they strengthen the bilateral character of the association between life satisfaction and BJW" (Correia et al. 2009, 1).

In fact, more happiness as life satisfaction and BJW are in a circularly supportive relationship, while more happiness as stronger positive affect and BJW are not in such a relationship. This indicates that BJW is in a circularly supportive relationships with happiness as a stable rather than a transient state.

But what is the relation among BJW, morality and happiness as a "trio"? Can BJW link morality and happiness? If there is such a link, this is how it would operate. If we are more moral, we might have a stronger proclivity toward BJW, while BJW might be conducive to happiness as life satisfaction. Conversely, if we are happier (having a higher level of life satisfaction) we might be more inclined to BJW and this inclination might make us more prone to act morally. Hence, according to this line of reasoning, BJW would explain, at least in part, the circularly supportive relationship between moral behavior and happiness (as life satisfaction).

The mediating role of BJW in the relationship between morality and happiness bears resemblances to the mediating role of BJW between belief in a free will and moral action (see Chap. 3). Empirical evidence suggests that the human belief in a free will is a significant motivation for them to act morally. Various empirical findings substantiate this: disbelief in free will decreases helpfulness and increases aggression (Baumeister et al. 2011), trust in free will has behavioral consequences, including increases in socially and culturally desirable acts (Baumeister et al. 2009), the readiness potential for acting is lowered in individuals induced to be skeptical about a free will (Rigoni 2012a), undermining free will can degrade self-control and can lead to other antisocial tendencies (Rigoni 2012b), mistrust in free will increases the tendency to cheat (Vohs and Schooler 2008).

In conclusion, it appears that BJW is an important disposition helping humans to believe in having a free will, to act morally and consequently increase their happiness (understood as overall life satisfaction).

5.2 Psychopathy

If morality and happiness are positively correlated, what can be said about the relationship between unhappiness and immorality? It has been demonstrated in a number of studies that they are also positively correlated (Love and Holder 2016). Morally

5.2 Psychopathy

inadequate people tend to have less emotional bonds with other people, while some of them might experience a feeling of guilt. All these feelings are not conducive to happiness.

Weak emotional bonds are characteristic for psychopaths. Psychopaths are often charming, but their interpersonal relationships are shallow. Even if the quantity of these relationships might be large, their quality is low. This should be expected in light of the fact that psychopaths also have a tendency to display hostility, impulsiveness and anger. Their emotional intelligence tends to be underdeveloped. Their romantic relationships are generally poor (Love and Holder 2016).

Hence, the results from various studies suggest that psychopaths are generally unhappy. They show low levels of positive emotions and life satisfaction, and high levels of negative emotions and depression. Love and Holder (2016) explains their unhappiness in part by the unsatisfactory quality of their romantic relationships.

Psychopathological immorality contributes to unhappiness, according to various findings. A reverse relationship has not been established (unhappiness contributing to psychopathological immorality). Hence, the two have not been proven to be in a circularly supportive relationship, as have morality and happiness.

Furthermore, psychopaths tend not to be aware of the gap between how humans act and how they believe they *ought* to act: the comprehension-motivation gap escapes them. Morally imperfect people (that is, all of us), or immoral people who are not psychopaths, are aware of it. Morally perfect people don't have a comprehension-motivation gap. Psychopaths are not aware of having it. Morally perfect people do not exist, psychopaths do. Hence, all people, psychopaths excluded, suffer to a smaller or larger degree from the comprehension-motivation gap. Only psychopaths are not aware of it. They lack the type of self-consciousness that is required for this awareness.

It is therefore possible to formulate the following classification of moral categories:

1. Moral perfection: always acting as one correctly thinks one *ought* to act. Such people do not exist.
2. Common (im)morality: occasionally acting in a manner that departs from what one deems to be morally right. A remedy to such behavior can be moral education and a strengthening of motivation. The latter can be addressed by MBE. It should be voluntary MBE.
3. Psychopathological im(morality): acting in a quantitatively and/or qualitatively excessive manner as one incorrectly thinks one ought to act. A remedy to this can also be MBE. As immoral psychopaths are unlikely to opt for voluntary MBE (they are not aware of the fact that they are immoral, as they don't truly understand what immorality is), compulsory MBE can be applied. In cases of dangerous psychopaths (e.g., certain incarcerated habitual offenders), compulsory MBE might be an urgent imperative. Moral education might also be helpful, however, as it can show the psychopath why her behavior is detrimental to her happiness. She will not understand what morality is, but might be sensitive to

the unhappiness immorality causes her. In the case of psychopaths, such moral education would frequently have to be compulsory.

In category 1 there is no comprehension-motivation gap. People who would fall into this category would always act in a morally right manner. But nobody falls into category 1. It is an empty set. In category 2 there is a comprehension-motivation gap. People act sometimes differently than they believe they ought to act—some more, others less. Most people fall to a higher or lower degree into category 2. In category 3 there is no comprehension-motivation gap, but for very different reasons than in category 1. In category 3 people don't understand that their behavior is sometimes morally wrong. They do not understand the difference between morally right and morally wrong. They do not comprehend morality. They are psychopaths.

5.3 Utilitarianism and Its Redefinition as the Maximization of Median Happiness

In what has been argued until now the role of happiness is obviously essential. It has been demonstrated that we ought to maximize happiness in order to enhance ourselves morally *and* that we ought to enhance ourselves morally in order to maximize our happiness. The point has also been made previously that the happiness we deal with here is not transient affects/superficial pleasures, but lasting and profound forms of happiness.

Apart from not being in a circularly supportive relationship with morality, the superficial pleasures have another important pitfall. If they are enjoyed in immoderation they can even lead to unhappiness. Stable and non-superficial types of happiness, sometimes coming close to *eudaimonia*, do not suffer from this pitfall. For example, excessive eating and drinking transforms pleasure into pain, while an abundance of reflection, enjoyment of art and similar activities do not lead to unhappiness (usually).

In some situations happiness is not a normal reaction. If you hear that your best friend has died in a car accident, a happy reaction would be inappropriate (and verge even on insanity). If you see someone who has caused major harm to you, a happy feeling and toothy smile when you see her would be inadequate. The first situation causes unhappiness, the second discomfort. But both reactions are transient. A lasting feeling of happiness can however still permeate us in painful situations or even in very distressing periods in our lives.

This brings my argument to the vicinity of utilitarianism. The utilitarian formula of maximization of happiness has to deal with several problems. One of them is the following: does maximization of happiness refer to average happiness or to total happiness[1]? Consider the following case. Imagine an influx of a large number of people from a country with low GDP to a country with high GDP. For example,

---

[1] I will abstain from references to the thoughts of Derek Parfit and others on this theme, as they would lead the discussion in a different direction and possibly violate the thematic unity of this book.

millions of people from Bangladesh enter Switzerland and commence their legal residence there. The Bangladeshi immigrants will benefit from living in a country with high GDP, while the Swiss will benefit from cheaper labor and lower prices. More importantly for our argument, in case the number of Bangladeshi immigrants exceeds the number of Swiss citizens, the Bangladeshi and the Swiss, taken together as a whole, will experience a net benefit. Still, the *average* material prosperity of the Swiss citizens will become lower, as the poor Bangladeshi immigrants will bring it down.[2]

Consider that maximization of happiness does not deal with average but with total happiness. Utilitarians would prefer 1000 almost perfectly happy humans to 500 perfectly happy humans. Similarly, they would prefer 6 billion rather unhappy humans to 6000 perfectly happy humans, because the little happiness in 6 billion humans contains more total happiness than the total perfect happiness of 6000 humans.

Hence, it is neither average happiness nor total happiness that seems to be able to survive as utilitarian models in these cases. The most promising approach is to look at the *median value* of happiness. The median value of happiness does not face the problems mentioned above. The median material prosperity of the Bangladeshi and Swiss would for example show a more accurate increase in material prosperity than average material prosperity of the Swiss and Bangladeshi would show. In this thought experiment, it is median material prosperity that best demonstrates whether material prosperity has been maximized. If we replace material prosperity with happiness, the results are similar. Maximization of happiness is in various cases best reflected by the maximization of the median value of happiness. Calculating median happiness is therefore sometimes the best way for utilitarians to calculate the maximization of happiness.

Bringing these arguments in line with the circularly supportive relationship between morality and happiness, the following can be concluded about humans in general and about humans as influential decision makers ("policy makers" in the broadest sense of its meaning):

1. Human individuals ought to maximize their happiness (at the same time not being detrimental to median happiness) in order to enhance themselves morally *and* human individuals ought to enhance themselves morally in order to maximize median happiness.
2. Humans as policy makers ought to maximize median happiness in order to enhance humans morally *and* humans as policy makers ought to enhance themselves morally in order to maximize median happiness.

Bartels and Pizarro (2011) notes that researchers have recently argued that utilitarianism is the appropriate framework by which to evaluate moral judgments. According to this line of reasoning, non-utilitarian solutions to moral dilemmas are morally inferior. Bartels and Pizarro (2011) present however the results of a study in which

---

[2]There are various other examples as well. If one accepts the average principle of utilitarianism, she can get to the point of justifying morally repugnant actions, such as killing disabled children.

participants responded to a battery of personality assessments, as well as to a set of dilemmas in which they had to opt for either utilitarian or non-utilitarian solutions. The study showed that participants who indicated a greater endorsement of utilitarian solutions had higher scores on dispositions of psychopathy, machiavellianism and life meaninglessness. In contrast to expectations following the assumption that utilitarianism is the appropriate framework for evaluating moral judgments, the study revealed that it is precisely those individuals accepting such a moral framework who tend to possess a set of psychological features that many would consider to be typically immoral.

Bartels and Pizarro (2011) offers cogent arguments in favor of the position that utilitarian views are not morally superior to other ethical views. But it would also be wrong to conclude the opposite: that utilitarian views are morally inferior to other ethical views. It would be especially troublesome to insist on the link between utilitarianism and psychopathy.

As has been previously argued, namely, the cognition-motivation gap does not exist in the minds of true psychopaths. They combine poor moral behavior and poor moral judgment. They are not aware that they act differently than is morally right. Psychopaths don't understand why something is morally right and something morally wrong. Hence, voluntary MBE is an ineffective approach to psychopaths. They don't understand that something ought to change in their moral judgments and behavior.

Utilitarians, on the other hand, have no moral impairment. They distinguish between morally right and morally wrong. The fact that their conclusions might sometimes be counter-intuitive, can be a consequence of the possible wrongness of our moral intuitions, or it can be a consequence of a wrong calculation of the maximization of happiness. As has been argued before, average happiness and total happiness might sometimes be fallacious indicators of maximum happiness. In the examples that have been offered, median happiness appears to be the correct calculation of maximum happiness in certain cases. If we calculate happiness in that way, utilitarianism might sometimes appear to be less morally counter-intuitive.

Finally, unlike psychopaths, utilitarians should be targets of moral education and voluntary MBE, as should be all other people who do not fall under category 3—the category of people who are unable to perceive the comprehension-motivation gap, who are unable to distinguish between morally right and morally wrong, who are psychopaths. Psychopaths are the only category of people for whom *compulsory* MBE might be justified—especially in case they are incarcerated habitual offenders who have committed one or more serious crimes.

References

Bartels Daniel, M., and David A. Pizarro. 2011. The mismeasure of morals: Antisocial personality traits predict utilitarian responses to moral dilemmas. *Cognition* 121: 154–161.

References

Baumeister, R.F., E.J. Masicampo, and C.N. DeWall. 2009. Prosocial benefits of feeling free: Disbelief in free will increases aggression and reduces helpfulness. *Personal. Soc. Psychol. Bull.* 35: 260–268.

Baumeister, R.F., A.W. Crescioni, and J.L. Alquist. 2011. Free will as advanced action control for human social life and culture. *Neuroethics* 4: 1–11.

Correia, Isabel, Mari Toscano Batista and Lima Maria Luisa. 2009. "Does the belief in a just world bring happiness? Causal relationships among belief in a just world, life satisfaction and mood. Aust. J. Psychol. 00 (0): 1–8. (file:///C:/Users/Vojin/Downloads/RI_19CorreiaBatistaLima2009.pdf).

Harari, Yuval. 2015. *Homo Deus*. London, UK: Harvill Secker.

Love, A.B., and M.D. Holder. 2016. Can romantic relationship quality mediate the relation between psychopathy and subjective well-being? *J Happiness Stud* 17: 2407–2429.

Rigoni, D., S. Kühn, G. Sartori, and M. Brass. 2012a. Inducing disbelief in free will alters brain correlates of preconscious motor preparation: The brain minds whether we believe in free will or not. *Psychol. Sci.* 22: 613–618.

Rigoni, D., S. Kühn, G. Gaudino, G. Sartori, and M. Brass. 2012b. Reducing self-control by weakening belief in free will. *Conscious. Cognit.* 21: 1482–1490.

Vohs, K.D., and J.W. Schooler. 2008. The value of believing in free will: encouraging a belief in determinism increases cheating. *Psychol. Sci.* 19: 49–54.

Chapter 6
The Morality of a Global State, the Existence of Morally Superior Beings and "Special Suicide"—Brief Conceptual Explanations

In this chapter a number of conceptual explanations will be introduced that will be elaborated on in the essential last chapter of the second part of this book. It is also useful to the reader to have an awareness of these conceptual clarifications in the chapters preceding the last chapter of the second part of this book.

6.1 The Morality of a Global State

Financial success generally increases happiness. Ethics is important in business and in the financial world in general for a number of reasons. First, ethics creates trust. Without trust, business deals and financial transactions become more difficult. They slow down the economy. How can a society advance properly if people cannot have confidence in their financial institutions? How can an economy grow if people distrust banks?

Second, regulation can be frequently bypassed in the financial world. Ethics, on the other hand, is essential for successful business. Ethics, its constitutive part being trust, cannot be bypassed in the financial world, if that world is to function successfully.

Third, financial institutions are essential for economic growth and development, no matter whether we look at countries with high or low GDP. Financial services ought to be guided by ethical principles that facilitate their effectiveness. If they are not guided by these principles, they will not be sufficiently effective. Ethical behaviour is therefore key to global financial services. Hence, ethical behavior in the financial world legitimizes wealth creation, which is a major driver of overall global development.

The economic crisis that started in the first decade of the 21st century had an adverse impact on trust in financial institutions. Fraudulent behavior that has been exposed showed that ethics in the financial world has been frequently missing, while its effects on trust in financial institutions had additionally adverse effects on world economies. Instead of financial success contributing to happiness and thus to morally enhanced behavior, it created greed in some people and fraudulent behavior to satisfy it.

Global poverty, on the other hand, significantly reduces happiness. Our spending on harmful or unnecessary things (fur, smoking etc.[1]) and the global inequality this exposes, together with national policy making based on narrow national interests, are major contributing factors to global unhappiness and thus global immorality.

A global state would be more moral than the current world in the following ways:

1. A global authority would be more capable of eliminating extreme poverty in the world and thus in increasing not only median, but also total and average material prosperity. This would have a positive effect on median, total and average happiness and therefore on global morality.
2. A global state would help humans understand that the individual and not a national, ethnic or other collectivity is the basic unit of morality. Individual morality would be boosted in a global state.
3. A global authority with a global legal system would be more effective in curbing unethical behavior by greedy individuals who control much of global wealth.
4. A global authority, if liberal and democratic, would be more effective in preventing the Harari scenario. It would be a barrier to morally irresponsible use of AI and big data.

6.2 The Existence of Morally Superior Beings

The existence of beings that are morally superior to humans has non-zero probability. If the universe is infinite, the existence of morally superior beings is certain. If the universe is finite but continuously expanding (what various contemporary physicists claim), the existence of morally superior beings is also certain. But do we understand these morally superior beings? If their superiority is based on superior cognitive abilities, the implication is that we don't understand them. It is not possible to understand those who are cognitively superior to us. At least not in all realms of cognition.

This has two major repercussions. First, the issue of theodicy issolved in a new manner. If we accept that we don't understand the morality of morally superior beings, we can infer that we don't understand morally superior notions of good and evil. Hence, what we perceive as evil in the world might very well be something that does not qualify as evil in morally superior notions of good and evil (see Part II of this book).

Second, the implication is also that we are vis-à-vis these morally superior beings in a status that might resemble the status of psychopaths to us. We don't understand why the moral conceptions of morally superior beings are superior to ours, similar

---

[1] http://blog.practicalethics.ox.ac.uk/2014/03/is-smoking-morally-wrong/; last accessed on 1 November, 2020.

to psychopaths not understanding why our moral conceptions are superior to theirs. We are morally impaired in relation to them, as are psychopaths in relation to us.

6.3 Introducing Special Suicide

A global state can hardly help humans understand that it is morally justified to gradually replace them by morally superior posthumans. For this to understand, a specific type of moral education is needed. Interestingly, it is Confucianism that can offer something that might be a hint: the conception of "altruistic suicide"; this type of suicide is required when humans cannot live "*Ren*", that is, a virtuous life.

An example of virtue in the form of "altruistic suicide" is the sacrifice of one's life for his children's lives. An analogous example would be the sacrifice of the human species for posthumans. Not much different from loving our children more than ourselves, we might love a morally superior species of posthumans more than existing humans. We don't have to sacrifice our individual lives for that. The sacrifice of the human species does not imply individual self-annihilation. It is not exactly Confucian "altruistic suicide". It is suicide of our species ("*special* suicide").

- "Special suicide" is a form of suicide in which we do not sacrifice *individual* lives or other individual interests but *special* (pertaining to species) lives or other *special* (pertaining to species) interests.

  These arguments will be expanded in the second part of the book.

Part II
Evil and Good

Chapter 7
Evil

It has been argued already that certain conceptions of the good are not geographically and/or historically determined. Some of these conceptions are the moral notions we have apparently been born with (see Chaps. 1 and 2 of Part 1). Those conceptions about which there is no consensus in (the vast majority of) humankind, are prone to being differently evaluated by deontologists, utilitarians, virtue ethicists; by all humans. A lot of evaluations about good and evil tend to focus on what is good. I will take a different route in this chapter, focusing to a large extent on evil.

Some atheists tend to suggest that God does not exist, but that "Evil" somehow does exist. Famous thinkers such as Richard Dawkins, Daniel Dennett or Christopher Hutchins declare themselves as atheists, but appear to believe in something resembling "absolute Evil". Hutchins: *"… many of the attempts to introduce "complexity" into the picture strike me as half-baked obfuscations or distractions… against the tendencies of euphemism and evasion, some stout simplicities deservedly remain… The regimes of Saddam Hussein and Kim Jong Il and Mahmoud Ahmadinejad fully deserve to be called 'evil'"*.[1]

Is the suggestion here that while God does not exist, Satan does exist? Is belief in Good and God superstition characterized by childish and naive character traits and/or lack of intelligence and education, while belief in Evil (Satan?) is scientifically founded?

But what is Evil? How can we have a conception of evil if we don't have a conception of good? Evil seems to be defined by good. The absence of the good, or the opposite of the good, or a denial of the good, are all conceptions of evil. Evil appears to be defined by the good. It seems to be parasitic on the good.[2]

In what follows a number of phenomena about which there is a broad consensus in much of humanity that they are mostly evil, will be discussed. In the course of

---

[1] https://www.litkicks.com/EthicsAndTheConceptOfEvil; last accessed on 1 November, 2020.
[2] Interestingly, Evil parasiting on the Good is also present in the conception of Lucifer as the Fallen Angel—someone who was good, but who rebelled against the Good (God) and thus became the carrier of Evil. Hence, Satan came into being as an entity parasiting on God.

the discussion of these phenomena, the following three options on how to face them will be briefly addressed:

1. Moral education—provided that it can effectively enhance the understanding of morality by (some) humans; it would help humans understand better what is morally right and wrong.
2. MBE—provided that it can enhance the motivation of humans to act as they understand is morally right to act.
3. Public deliberation—provided that a society has not succeeded to arrive at a consensus on what is right and wrong when certain issues are concerned.

7.1 Deprivation of the Greatest Good

Murder[3]

> Do not murder. (Ancient Jewish. Exodus 20:13)
>
> In Nastrond (= Hell) I saw... murderers. (Old Norse. Volospd 38, 39)
>
> I have not slain men. (Ancient Egyptian. From the Confession of the Righteous Soul, 'Book of the Dead', v. Encyclopedia of Religion and Ethics f = ERE], vol. V, p. 478)

The moral wrongness of (most) murder(s) can have diverse groundworks, most of them based in different approaches to the origin of morality, including theological, social contract, intuitive morality. In all these approaches murder deprives the victim of something good: life—life as a sacred gift, life guaranteed by a promise to preserve it (grounded in the social contract) or life guaranteed by the confidence that the intuition not to murder is a moral value that is also cherished by the murderer.

This discussion can follow the direction of merciful killing (euthanasia—preferring the value of alleviating major suffering to the value of life). Nonetheless, killing remains a deprivation of some good. So does murder.

Killing—euthanasia

The majority of "murders" are in all societies immoral. They have a malign aforethought. "Killing" is a broader category and consequently has a different status. It is not associated with a malign aforethought. Although killing entails a serious deprivation of some good (life), in certain cases it is even morally allowed. In case that it is outweighed by a more valuable good (the end of suffering), it might even be a moral duty. That can be the case in euthanasia.

Sinnott-Armstrong and Miller (2013) discusses the possibility that killing is wrong because it deprives the one who is killed of all remaining abilities. The article poses

---

[3]The thoughts on good and evil are cited from the Appendix to the *Abolition of Man*, by Lewis (1943).

7.1 Deprivation of the Greatest Good   47

the question whose ultimate goal is to justify organ donation after cardiac death—where the patient has brain damage and is on life support, but he isn't dead because the heart could start beating again. The authors give two scenarios of 'Betty' who is shot in the head by 'Abe'. In the first scenario, she is killed. It would be all right to harvest her organs in this case, because she is dead. In the second scenario, she is not dead, but totally disabled—she is conscious but totally unable to control any of her actions, or even to process thoughts or experiences. The authors argue that 'Betty' has already had her ability to experience taken away, so it is not immoral to kill her in order to harvest her organs.

They argue that there is practically no difference between causing death and causing total disability, because both outcomes result in her losing the ability to experience a pleasurable life. According to their argument, what's immoral about either is that it causes the total and irreversible loss of abilities. It is therefore not necessarily immoral to take the life of someone who has already experienced this total and irreversible loss, because there's nothing left to take from her (Sinnott-Armstrong and Miller 2013: 3).

The authors develop the notion that killing for the sake of alleviating severe suffering is not the only morally justified reason for euthanasia. If someone is permanently deprived of her abilities, euthanasia is morally allowed as well. It has to be noted, however, that wrong diagnoses of permanent disability have contributed to a change of the diagnosis "permanent vegetative state" to "chronic vegetative state". Hence, caution is advised when predicting that a patient will not be able to recover. Only if such a prediction is well-founded, is the argument of Sinnott-Armstrong and Miller valid. In that case, euthanasia may also be morally justified as the patient is permanently deprived of her abilities.

Killing that is not Euthanasia

> By the fundamental Law of Nature Man [is] to be preserved as much as possible. (Locke, Treatises of Civil Govt. ii. 3)
>
> The killing of the women and more especially of the young boys and girls who are to go to make up the future strength of the people, is the saddest part… and we feel it very sorely. (Redskin. Account of the Battle of Wounded Knee. ERE)

In most societies it is morally allowed:

- to kill in self-defense;
- to kill for the sake of saving the lives of one's family members or close friends (or possibly of those who cannot defend themselves[4]) who are threatened to be murdered;

---

[4]Not only killing, but violence in general is frequently morally justified in self-defense (this applies also to states which defend their sovereignty if being attacked) and in defense of the weaker, that is, of those who cannot defend themselves successfully. The latter does not apply to states, because

- to kill in order to save the life of any innocent human who is threatened to be murdered and who cannot defend herself successfully;
- to kill in wars (without committing war crimes);
- to kill an animal of lower moral status than humans have (mostly flies, rats or animals humans like to consume).

I did not list here capital punishment, because the opinions on that issue are very divided. The basic argument behind retribution and punishment is:

- all guilty people deserve to be punished;
- only guilty people deserve to be punished;
- guilty people deserve to be punished in proportion to the severity of their crime.

This argument requires people to suffer in a way appropriate to their crime. Each criminal should get what her crime deserves. In the case of a murderer the deserved punishment seems to be death.

On the other hand, various arguments can be given against capital punishment. The least important one is certainly not that someone may be executed for a crime she did not commit. When the real culprit is discovered, it is sometime too late: the innocent person has been executed already.[5]

Another type of killing that I did not list is suicide. Suicide is also an issue that is marked by disagreements. Although the respect of one's freedom entails her right to decide about her life, objections to suicide include:

- The argument in many religions that any murder is a sin, including the murder of oneself. In fact, in a certain sense suicide might be the worst of all sins, as it makes repentance impossible. Once you are dead, you cannot repent. You committed a crime without the possibility of penitence.
- Some people want to commit suicide although they have the possibility to have a decent life they can be satisfied with. The solution in their case is not suicide but therapy.

A third type of possible killing that I will not discuss is induced abortion. Not only that opinions on this issue are very divergent throughout the world, but the definition of induced abortion as "killing" is debated as well. Some people do consider induced abortion to be killing, even murder, while others do not perceive it in that manner.

The morality of killing in war is a very specific case. Pacifists deny that killing in wars is morally allowed. However, irrespective of being a pacifist or not, the perspective can be taken of a soldier who is drafted and who can be killed by the enemy or kill the enemy instead. He can also refuse to take orders, which in war

---

sometimes defense of a weaker state is unjust, whereas the human cost may also be too high if a weaker state is being defended by a stronger state.

[5]There may be an analogy here with euthanizing people in a "permanent vegetative state". The development of sophisticated fMRIs has shown that the vegetative state does not have to be merely "vegetative", but more than that. Similarly, DNK analyses have demonstrated that some sentenced and executed prisoners were innocent. Modern technologies can therefore be of help in morally justified decision making both in medicine and in jurisprudence.

might be punished by death. Hence, killing in war can be considered as a type of self-defense: if you don't kill, you might be killed by the enemy or you might be executed for refusing orders.

But are there cases in which *murder* is morally allowed? What about a tyrant who is conducting a campaign of terror against her subjects? Is the murder of such a person morally allowed? We can also think of blood revenge. This type of revenge is morally allowed in some primordial societies.

It can be argued that, from a moral standpoint, the killing of a tyrant is not an immoral murder, but a sort of self-defense, or rather the defense of a nation/ethnic group or another community. If this community is threatened by the tyrant and the threat is existential (mass murders of members of a community) killing the tyrant amounts to self-defense.

The point can even be made that blood revenge is also a type of self defense. For a family whose member has been killed by the member of another family, revenge is a deterrent for the culprit family not to commit another crime. Nevertheless, the argument that blood revenge is moral, is wrong. First, in a society with a functional legal system, blood revenge is unnecessary and immoral. Second, even if there is no functional legal system in a country, the concept of punishing someone who committed a serious crime on a person from another family, should not be based on taking revenge on a person the wrongdoer loves. Blood revenge is however based precisely on this concept. Here is an example:

Person x from family A kills person y from family B. The brother of person y from family B kills then not person x from family A, but his brother z. The pain that has been inflicted on person x (the initial killer) by the murder of his brother is then caused by the killing of an innocent person z (the brother of x).

The murder of an innocent person is however immoral *in principle*. Hence, blood revenge is immoral *in principle*.

Why is it morally wrong, evil, to kill innocent people? The Evil is contained in the fact that it is based on deprivation of some Good. This Good is life. Life is good. Preserving life is good. Exceptions are often considered to be euthanasia in certain cases, killing in self-defense, killing for the sake of saving the lives of one's family members or close friends who are threatened to be murdered, killing in order to save the lives of innocent people who are threatened to be murdered and who cannot defend themselves, killing in wars (without committing war crimes), killing animals of lower moral status than humans.

It is worth noting that there is a rather wide trans-cultural consensus on these issues. Opinions are still strongly divided in the cases of capital punishment, induced abortion, suicide and some forms of euthanasia, but in most other cases even culturally very different societies agree that life is a Good, that depriving someone of her life is Evil, and on how life should be protected by individual action. Cultural practices that can be considered as immoral (blood revenge) often have only limited numbers of followers.

How can MBE and moral education be of help here? MBE can be useful in all cases in which people are aware of the values that have been discussed in this section, but fail to act in line with this awareness. They lack motivation to act in line with how they understand is morally right to act. Moral education can be applied in cases in which a rather small group of people fails to understand that their practices are morally wrong (e.g., blood revenge). All issues regarding which there is broad disagreement (e.g., capital punishment, suicide, induced abortion or certain forms of euthanasia) require moral reflection based on broad public deliberation. They can only be addressed by such an approach, not by traditional moral education or MBE.

7.2 Deprivation of Lesser Goods

Depriving a human of his life is a major and permanent deprivation. Depriving him of some of his property or the truth is a lesser and frequently not permanent deprivation.

Stealing

> Thou shalt not steal. (Ancient Jewish. Exodus 20:15)
>
> I have not stolen. (Ancient Egyptian. Confession of the Righteous Soul. EREy. 478)
>
> Has he drawn false boundaries? (Babylonian. List of Sins. ERE v. 446)
>
> Choose loss rather than shameful gains. (Greek. Chilon Fr. 10. Diels)
>
> To wrong, to rob, to cause to be robbed. (Babylonian. Ibid.)
>
> Justice is the settled and permanent intention of rendering to each man his rights. (Roman. Justinian, Institutions., I. i)
>
> If the native made a "find" of any kind (e.g., a honey tree) and marked it, it was thereafter safe for him, as far as his own tribesmen were concerned, no matter how long he left it. (Australian Aborigines. ERE v. 441)
>
> Whoso takes no bribe… well pleasing is this to Samas. (Babylonian. ERE V. 445)

Similar to killing, stealing is mostly considered as morally wrong. We can come up with a huge number of examples of stealing that are obviously morally wrong: stealing a lady's purse in the subway, charging your customers more by offering them fake prices, issuing fake invoices in order to pocket the difference with the price for which something has been sold….

There are however instances of stealing that a respectable number of people justify. Stealing food if you don't have the means to feed yourself and your family might be morally right. Stealing back your money from a thief who has stolen it from you might be the right thing to do in certain situations. On the other side, stealing from a thief who has not stolen from you is morally wrong, because that what you steal from the thief does not belong to you either.

What about stealing from rich companies in order to achieve a less unjust distribution of wealth? If a company created jobs and allowed the poor to work and for the salaries they earn to purchase goods, then, assuming just working conditions and compensation, there is no moral justification to steal from such a company. On the other hand, if the company had obtained its money by robbing the poor, then simply giving the money back to the poor could be considered as distributive justice. Stealing it back might however open the door to lawlessness (although the company might reside in a country that is already lawless).

Where can MBE and moral education step in? MBE can deal with those cases in which stealing is obviously wrong. In such cases people who steal know that it is wrong, but still do it. Their behavior exposes the comprehension-motivation gap. In cases in which certain people don't understand that stealing is wrong (e.g., stealing from a thief who has not stolen from you), it is moral education that can help. In cases in which stealing is not obviously wrong (the example of companies who robbed the poor), public deliberation is also an approach that might be useful in order to assess the morality of stealing from a robber.

Lying

> Slander not. (Babylonian. Hymn to Samas. ERE v. 445)
>
> With his mouth was he full of Yea, in his heart full of Nay? (Babylonian. ERE V. 446)
>
> I have not spoken falsehood. (Ancient Egyptian. Confession of the Righteous Soul ERE v. 478)
>
> A sacrifice is obliterated by a lie and the merit of alms by an act of fraud. (Hindu. Janet, i. 6)
>
> Whose mouth, full of lying, avails not before thee: thou burnest their utterance. (Babylonian. Hymn to Samas. ERE v. 445)
>
> Thou shalt not bear false witness against thy neighbour. (Ancient Jewish. Exodus 20:16)
>
> Hateful to me as are the gates of Hades is that man who says one thing, and hides another in his heart. (Greek. Homer. Iliad, ix. 312)
>
> The foundation of justice is good faith. (Roman. Cicero, De Off. i.vii)
>
> I sought no trickery, nor swore false oaths. (Anglo-Saxon. Beowulf, 2738)
>
> The Master said. Be of unwavering good faith. (Ancient Chinese. Analects, viii. 13)
>
> A sacrifice is obliterated by a lie and the merit of alms by an act of fraud. (Hindu. Janet, i. 6)
>
> Whose mouth, full of lying, avails not before thee: thou burnest their utterance. (Babylonian. Hymn to Samas. ERE v. 445)
>
> I have not spoken falsehood. (Ancient Egyptian. Confession of the Righteous Soul ERE v. 478)
>
> I sought no trickery, nor swore false oaths. (Anglo-Saxon. Beowulf, 2738)

> The Master said. Be of unwavering good faith. (Ancient Chinese. Analects, viii. 13)
>
> In Nastrond (= Hell) I saw the perjurers. (Old Norse. Volospd 39)
>
> Hateful to me as are the gates of Hades is that man who says one thing, and hides another in his heart. (Greek. Homer. Iliad, ix. 312)
>
> The foundation of justice is good faith. (Roman. Cicero, De Off. i.vii)
>
> [The gentleman] must learn to be faithful to his superiors and to keep promises. (Ancient Chinese. Analects, i. 8)
>
> Anything is better than treachery. (Old Norse. Hdvamdl 124)
>
> Thou shall not bear false witness against thy neighbour. (Ancient Jewish. Exodus 20:16)

According to some scholars, deception is a physiological manifestation of the need of organisms to survive. Deception has a similar role as speed, sharp teeth, sharp and strong claws. Spiders, when in danger, fall motionless in order to simulate death. Some crabs, when attacked, throw away their claws which keep moving while the crab itself disappears. Some birds imitate the sound of snakes, chameleons adopt the color of the environment. There are plenty of other examples of deception in nature (Campbell 2002).[6] These deceptions have all the elements of a lie, except that they are not verbalized.

Lying and deception is also natural for primitive humans as is deception for animals.[7] With the development of civilization, it might however be argued, lying has become to be considered as a manifestation of weakness and immorality. Hence, stronger human organisms might be considered to deceit less. A weaker proclivity to lie might therefore often be a manifestation of both strength and morality.[8]

Campbell (2002) relativizes truth, however. He considers lying frequently as a means to arrive at some "deeper" truth. In his view, lying develops with the development of civilization. It is not morally abhorrent, but laudable.

In marketing, lying and deceit is not allowed. But "puffery" is. For example, one cannot advertize cigarettes with the argument that they benefit the consumer's health, but she can advertize a car by claiming that when driving it you feel like God. The advertiser does in this case not really mislead the potential buyer, but expresses a subjective notion. Moreover, hardly anyone would believe that by driving any car he would feel like God. The advertiser engages in a type of exaggeration ("puffery") bordering a lie, but not being a real lie.

Christian teachings consider lying a sin. It is mostly a venial sin, but sometimes it is a mortal sin. For example, lying under oath with the awareness that an innocent

---

[6] See also https://www.ncbi.nlm.nih.gov/pmc/articles/PMC1068309/pdf/jnpsycho00045-0026.pdf; last accessed on 1 November, 2020.

[7] "Primitive" does not denote something pejorative here. Its meaning refers to an early stage in an entity's evolutionary or historical development.

[8] https://www.ncbi.nlm.nih.gov/pmc/articles/PMC1068309/pdf/jnpsycho00045-0026.pdf; last accessed on 1 November, 2020.

human will be convicted on the basis of your false testimony and sentenced to death or life inprisonment, is a mortal sin. In fact, we deal here with the Ninth Commandment of the Decalogue: "Thou shalt not bear false witness." In Christian dogma lying is not allowed under any circumstance. Since lying is inherently illicit, it cannot be made licit because of specific circumstances.[9]

In the case of lying, MBE can help in instances in which someone knows that it is morally wrong to utter a lie in a certain situation, but still does it, frequently because of a perceived benefit she expects to gain with her lie. Moral education or even therapy can be applicable when people lie because of prejudices or low self-esteem (e.g., lying about your level of education, age or sexual performance). Public deliberation might be useful in those cases in which it is not obvious that one should or should not lie (e.g., lying that your child does not have a minor medical condition if the truth could lead to him being teased in school).

7.3 Is "Evil" Always Evil?: Sexual Adultery, Prostitution, Sugaring, Paraphilias

Adultery

> Thou shalt not commit adultery. (Ancient Jewish. Exodus 20:14)
>
> Has he approached his neighbour's wife? (Babylonian. List of Sins. ERE v. 446)
>
> I saw in Nastrond (= Hell)... beguilers of others' wives. (Old Norse. Volospd 38, 39)

Adultery is extramarital sex. It is considered as morally wrong in most societies. As it is related to other sexual interactions, I will compare and contrast these interactions in one single section.

Different ethical theories take different views on adultery. The Decalogue considers adultery a sin, but a utilitarian might reason as follows: does the pain that Greg causes to his spouse outweigh his happiness, the happiness of his paramour and the pain of her husband? If not, adultery might be morally justified in this case. But also some deontologists might reason that a wife's duty of marital fidelity no longer stands if her husband does not love her, touch her, talk to her, but does pay regular visits to his mistress.

Hence, adultery is an activity that can be addressed by moral education or MBE. This would be the reasonable course of action if someone cheats on his spouse just because he is motivated to have sex with as many good looking women as possible

---

[9]Kant's position on lying bears similarities with the Christian viewpoint. The duty not to lie is a categorical imperative. No specific situation can morally justify lying. If it could, the duty not to lie would be a hypothetical rather than a categorical imperative.

(regarding them merely as a sort of "trophies"). This kind of person might be in need of both moral education and appropriate motivation. In certain cases, however, utilitarians, deontologists as well as advocates of other ethical theories, may have differing views on the morality of adultery. Public deliberation among ethicists and the public at large would be a reasonable course to be taken by a society in such cases.

The most obvious case when adultery is morally right (certain religious views I will disregard here) is when the spouses agree to have extra-marital sex. If they don't have children or if it doesn't hurt their children (they might not know it), adultery may be morally justified by many ethicists. Does this imply that prostitution is morally right as well, provided that it is performed by consenting adults?

Prostitution[10]

As prostitution is the exchange of sexual activity for payment, its moral dilemmas can reside in sexual activity, payment and exchange. Sexual activity, payment and exchange are in themselves however morally uncontroversial. Hence, it is only the combination of the three that can raise moral dilemmas.

Sexual activity involves various forms of exchange. As sexual activity is not being considered as immoral, sexual exchanges shouldn't be either. It is then the exchange of sexual activity *for payment* that can theoretically be regarded as immoral.

What can be wrong with exchanging sexual activity for payment?:

1. Payment may be withheld. Prostitution can take a variety of forms. It can consist of, for example, sexual activities within a brothel, it can consist of the company of a prostitute (escort) at parties, vacations, anniversaries, dinners (not necessarily combined with sex), striptease or "erotic massage". But all of them have one feature in common. They are exchanged for payment.

    If a prostitute has sex against her will and her payment is being withheld, it is possible to talk about "sex slavery". Sex slavery is however not prostitution. A slave can be forced to perform various activities, including sexual activities, but (s)he is primarily not a prostitute but a slave. Unlike a prostitute who receives payments for her activities, a slave doesn't receive payments for anything, including sexual services. Hence, it might be argued, cogently, that sex slavery is a form of rape. It is certainly not prostitution. Being enslaved means that the slave does not have the freedom to leave his/her job. Prostitutes generally do have such a freedom.

2. STDs may be spread by prostitution. This is possible as prostitutes are promiscuous in most cases. But promiscuity does not imply an increase in STDs if promiscuous individuals practice safe sex and are regularly tested for STDs. Mandatory testing for STDs would be an excellent program benefitting prostitutes. Furthermore, not only that it would protect them and their clients, but it would also be a source of revenue for the state.

3. Prostitutes are being "exploited", as they do not enjoy their job. Moreover, they don't prostitute themselves on a *genuinely* voluntary basis.

---

[10]The following paragraphs are derived from Rakić (2020).

This is not a cogent argument. First, a lot of employees don't enjoy their jobs. That does not necessarily mean that they are being "exploited". Second, the prostitute, even if she does not enjoy sex with her client, may enjoy the fact that her sexual services will be compensated by payment. She can enjoy her sexual activity because she knows that its result will be enjoyable.

4. In selling her body for sex, the prostitute's body is being treated as a commodity. This is also not a persuasive argument. First, a lot of people sell their bodily work. The argument that they do not sell their bodies, but use them to perform work, a certain skill or technique, is also refutable. Prostitutes, namely, also do not simply "sell" their bodies. They use them to perform certain skills or techniques. They also work. Second, it might be argued that models who advertise make-up treat their bodies as commodities, as they do not perform certain skills. In the case of advertising make-up, the model hardly needs to have any skill. Nonetheless, advertising make-up is not considered to be a morally dubious activity. Third, people already legally sell their body parts: blood, sperm, eggs, plasma or hair. In the case of surrogate motherhood, they also rent body parts (the uterus in the first place). If all the above mentioned activities are not immoral or outlawed, there is no reason prostitution should.

5. Prostitutes are being used as means to an end. Along Kantian lines, it might be argued that using someone merely as a means to an end violates the principle of humanity.

   However, we use people on a regular basis in ways that allegedly violates the principle of humanity. For example, I order a wine and use the waitress to bring me the beverage. Am I violating the principle of humanity in that way? No, because she gets paid (and possible tipped) for bringing me the wine.

   The same applies to prostitutes. The client has sex with a prostitute, she performs pleasurable activities on him, and the client pays (and possibly tips) the prostitute for the pleasure he has enjoyed. It is not a violation of the prostitute's humanity because it is a trade that occurs between two consenting individuals for mutual benefit. In general, whenever there is an exchange of a service for a payment between consenting individuals for mutual benefit, the service provider is not being used as a means to an end in a way that might violate the principle of humanity.

   In conclusion, the question that has been raised about the morality of prostitution can be answered as follows: prostitution performed by consenting adults is largely morally right. The general population is in need of moral education in that regard, as the public attitude toward prostitutes is largely imbued with prejudices about the alleged immorality of the jobs they perform. Public deliberation is needed to raise awareness about this issue. Additional remarks about the ethics of prostitution can be found in the following section on sugaring.

Sugaring

As in the case of other forms of prostitution, consent is essential for the exchange between sugar babies and sugar daddies. But unlike regular prostitution, in this type of exchange the category of informed consent is essential. If the sugar baby gives the sugar daddy truthful information on her emotional bond (or lack thereof) with him (or vice versa) the exchange is morally justified. If not, we deal with cheating, exploitation, fake exchange, the use of people as means to an end and other types of degrading and immoral behavior. On the other side, in the case of genuine exchange between sugar babies and sugar daddies, we deal with a regular, moral form of prostitution.

Unlike prostitution that should acquire the status of a morally apposite profession, sugaring based on cheating should be criminalized. In case that such type of exchange is discovered, the sugar baby (and in some cases the sugar daddy or mommy) should be charged with committing a felony.

Nonetheless, the prospects for sugaring aren't that dreary. Modern technologies have also here a potentially ethical role, as they can develop sugaring practices that are based on informed consent without one party being double-crossed in the transaction. On online sugaring platforms one can register as a sugar baby or sugar daddy and specify the terms of the transaction.[11] In that case, sugaring is a type of prostitution, that is a consensual activity, and hence loses its morally abhorrent status.

Four vignettes follow:

1. Mr. Klaas Janssen is a successful businessman. He is 69 years old and a millionaire. Susanne (25) is considered by most men and women as very attractive. She works as Mr. Janssen's secretary. She is determined to marry him and one day inherit his wealth. Mr. Janssen has no children. Susanne is not explicit in her intentions, but Mr. Jannssen is nevertheless aware of them. Susanne knows that he is. Mr. Janssen and Susanne marry. Susanne gives birth to a child, Klaas jr. For 19 years they live a rather happy live. After Mr. Janssen's death, Susanne and Klaas jr. inherit Mr. Janssen's wealth.

2. Dr. Alan Smith (35) is a divorced university professor, adviser to the government, an established social scientist with an excellent career. He is a home owner and has no financial problems. Irma (21) is a rural beauty with ambitions to work at the university (like Alan) and to advise the government (like Alan). Alan falls in love with Irma. His suspicions that she is a sugar baby are diluted by the fact that she was an excellent student and that she seems to have the potential to build an academic career. Irma moves into Alan's home. After a few weeks Irma approaches the Prime Minister, suggesting that she might trade sex in exchange for a position in his office. The Prime Minister tries to limit the transaction to sex only, but Irma makes clear that only a full transaction is acceptable to her. The potential "romance" with the Prime Minister doesn't succeed. Irma switches to opening Alan's Will (one month after they met) and asks him to include her in it. Alan refuses. Irma doesn't give up. She insists on expensive hotels, restaurants, cheats Alan on finances, but also continues to be determined to build a university career. She is not interested in having children. Alan is, and breaks up with Irma. But Irma is persistent. She returns to Alan, at once expressing

---

[11] See: https://sugarbabies.co/about/sugar-babies/ or https://www.sugardaddy.world/sugar-daddy-websites; last accessed on 5 January, 2019.

the wish to have children. She presses Alan for a month. Alan gives in. After some time Alan and Irma get a daughter Sarah. Alan and Irma's mother care for the child, so that Irma has the opportunity to pursue her career. As the child develops a strong bond with her father and not with Irma, while Alan continues to be successful in his career, Irma becomes jealous and hateful both to Alan and to her child. She begins to physically abuse Sarah. Alan files for divorce. Irma does not want a divorce. As Alan insists, she has no choice but to accept it. Irma's strong political connections (built during her sugaring career), persuades the public prosecutor to drop the charges of child abuse against Irma, while the judge accuses the father of alienating Irma from her and Alan's child. Moreover, Irma decides to file charges against Alan's alleged psychological abuse of Irma. Consequently, Alan is prohibited from seeing Sarah, the child and Alan suffer emotionally, but Irma seems to enjoy to see Alan and her daughter suffer. The corrupted judge even appoints, at Irma's request, a team of psychiatrists who falsely diagnose Alan with paranoia. Alan's career suffers profoundly, but his daughter Sarah remains close to him. With fifteen years of age she moves to Alan's home and aborts all communication with Irma.

3. Mr. Jack Wilson is 70 years old and a millionaire. His wife has died and Mr. Wilson becomes interested in having a relationship with a much younger woman, preferably in her twenties. Jill is 23 years old. She was a good student and wishes to enroll into a postgraduate astro-biology program at Princeton University. She comes from a poor family, does not obtain a stipend, and her tuition cannot be waived. Nonetheless, her dream remains to graduate astro-biology at Princeton. Jill visits a sugaring site, registers as a sugar baby, meets Mr. Wilson who registered as a sugar daddy. They arrange the conditions for the transaction they wish to have. Jill will move into Wilson's home, have regular sex with him and agree with other women joining their sexual activities whenever it pleases Wilson. Moreover, Jill will accompany Wilson at parties, dinners and vacations. In return, Wilson will take full care of financing Jill's graduate studies in astro-biology, finance everything that the household needs (Jill will do the purchasing) and pay in full for Jill's clothing, her study trips, as well as their common vacations.

4. Jack Levy is a well-known writer. His books sell well and with the passing of years Jack becomes very wealthy. He is 55 years old. Florence is 25 and studies art. She meets Jack and falls in love with him. Florence finds Jack charming, is impressed with his creativity and brightness, as well as his ability to make money. As Jack has also fallen in love with Florence, the two start living together and marry a few years later. Florence gives birth to two children.

Vignette 1 has the attributes of a consensual transaction. It resembles therefore regular prostitution and appears therefore moral. As the relationship is not based on explicit informed consent, this type of sugaring might raise certain moral dilemmas—dilemmas prostitution never raises. However, Mr. Klaassen and Susanne develop a relationship that might have acquired emotional features with the passing of time. In that sense, it has become something more fulfilling than prostitution usually is.

Moreover, a child was born out of that relationship, which generally does not happen in prostitution-based transactions. Having developed emotions to each other and having given life to a child, Klaassen and Sussanne have elevated their relationship into something that is morally superior to prostitution.

Vignette 2 is also a consensual transaction, but one that is based on cheating. Irma is a double-crosser and profoundly immoral. Her university career and social status protect her from being treated as human vermin in the society she lives in. Irma's sugaring is in any respect less moral than prostitution is.

Vignette 3 is an example of "pure" sugaring. It is consensual and marked by full disclosure of the reasons of the transaction. The transaction is obviously morally superior to the case from Vignette 2, but also to the case described in Vignette 1. In Vignette 1 a happy, perhaps even emotional, relationship develops and a child is being born (neither taking place in the case of Vignette 3). Nonetheless, the transaction itself is in Vignette 3 morally superior to the one in Vignette 1, because its terms are precisely specified. In Vignette 1 some suspicion is being left as to the motives of the transaction between Klaassen and Susanne. In Vignette 3 everything is clear.

Vignette 4 does not deal with sugaring. In some individuals the very idea of a beautiful young woman having an older or middle-aged partner raises eyebrows, but they should be careful in their judgment. Levy and Florence are in a truly loving relationship. The exchange taking place between them is primarily love.

Conclusion 1: If we consider the morality of the sugaring transactions taking place, the taxonomy of the vignettes is, in decreasing order of morality: Vignette 3, Vignette 1, Vignette 2. However, if we look not merely at the transaction, but also at the relationship that has been built on the transaction, the sugaring taxonomy of the vignettes may be, in decreasing order of moral adequacy: Vignette 1, Vignette 3, Vignette 2. Vignette 4 is not a case of sugaring and is therefore excluded from the sugaring taxonomy.

Conclusion 2: MBE and moral education can be of help to motivate potential sugar babies (and sugar daddies/mommies) to abstain from sugaring, unless they engage in a transparent transaction. In the case of prostitution, moral education and the enhancement of moral motivation by MBE would be ineffective, as in the case of prostitution it is not widely accepted how (im)moral it is and which values that pertain to prostitution to defend. The complexities of various types of prostitution and sugaring make them good candidates for being subjected to moral judgments on the basis of public deliberation.

Paraphilias

A paraphilia is the experience of sexual arousal caused by unusual organisms (that is, unusual for human sexual attraction), individuals, objects or situations. It is a disorder provided that it causes pain/distress to the person having it, that it is bad for her inclusive fitness and/or that it is immoral. The only aspect of paraphilias that are of interest to my argument is whether they are immoral.

There are several hundreds of paraphilias, some of them highly unusual and for many people sickening. That does not have to mean that they are immoral. As long as a sexual activity is consensual and as long as it does not cause third parties to suffer, it is not an immoral activity. A few examples follow. Mechanophilia (sexual arousal by cars or machines) is a weird sexual disorder, but there is nothing immoral in it if it does not cause distress to others. Similarly, auto-haemofetishism (bleeding oneself; sometimes called "auto-vampirism") is undoubtedly an unusual, sometimes dangerous sexual disorder. But it is not an immoral disorder if it does not cause stress to third parties (e.g., if it is being performed only in the presence of consenting individuals) and, possibly, if it is not fatal to the person who performs it. On the other hand, exhibiotionism (exposing one's genitals to non-consenting individuals) or frotteurism (rubbing against a non-consenting individual) are immoral because they are directed at non-consenting individuals. Sex of adults with prepubescent children (paedophilia) is immoral because a child of that age, even if giving "consent", is incapable of giving *informed* consent. It is therefore coercion (frequently with highly traumatic consequences for the child).

Further, some paraphilias have not been classified as disorders in any DSM Manual. If they are a disorder, however, paraphilias can be, but don't have to be immoral sexual disorders. Hence, paraphilias are not always immoral. The (informed) consent issue and the distress to third parties issue are the essential elements for determining the immorality of a sexual activity. Those paraphilias that include informed consent and an absence of distress to third parties are not immoral, no matter how weird or sickening they might be (see Quinsly 2011). A further discussion follows of several controversial paraphilias and sexual activities that are not necessarily paraphilias.

Rape

There are immoral sexual activities that are not necessarily paraphilias. It is possible to argue that rape (of an adult human being) is one of them, at least in certain cases. It may not necessarily be a sexual disorder, even not a paraphilia, but rather a moral disorder. If a man has sex with an adult woman, there is normally nothing wrong with him. But if he rapes her, he has an utter lack of empathy which qualifies him as cruel and immoral.

There are rare cases however in which rape might be justified. For example, if terrorists or other deranged criminals ask someone to rape his wife (child) in order for them to spare the lives of all his family members—rape might be morally justified. The reason is that being raped is mostly considered as a lesser evil than being murdered.

Although in the above example the victim could be morally justified for being forced to rape another victim, the ones who demand the rape are of course doing something immoral. Their action can be qualified as sadistic torture. Although sadistic torture is immoral (unless consensual), sadism itself (defined narrowly, as a *sexual* disorder), similar to many other sexual deviances, does not have to be psychopathy. The reason is that psychopaths tend to be indifferent toward much of their immoral behavior: they don't understand what is wrong with it, but also don't tend to enjoy it. Sadists, on the other hand, enjoy their disorder.

Torture

> He who is cruel and calumnious has the character of a cat. (Hindu. Laws of Manu. Janet, Histoire de la Science Politique, vol. i, p. 6)
>
> Terrify not men or God will terrify thee. (Ancient Egyptian. Precepts of Ptahhetep. H. R. Yi2i\\, Ancient History of the Near East, p. i3}n)
>
> I have not brought misery upon my fellows. I have not made the beginning of every day laborious in the sight of him who worked for me. (Ancient Egyptian. Confession of the Righteous Soul. ERE v. 478)

What about torture that has a purpose other than the enjoyment of the torturer? Can it be morally right? A utilitarian reasoning may be that when the overall outcome of lives saved due to torture is positive, torture may be morally justified. Deontologists, on the other hand, respect general rules and values (duties) that ought to be upheld regardless of the outcome. If the outcome of torture is uncertain or if the outcome cannot be traced back to the use of torture, some utilitarians may also consider torture to be morally wrong.

Consider the following case. Torture aimed at extracting confessions is morally unacceptable and illegal in all European countries for at least 168 years (Portugal and the Swiss canton Glarus being the last country\entity to abolish torture in 1838 and 1851 respectively). Confessions obtained by torture are not only morally inappropriate from the perspective of humaneness, but they have significant limitations. One of them is that tortured people may easily confess to having committed a crime they have not committed, in order to be spared of further torture. In such cases the outcome of torture is uncertain or the outcome cannot be traced back to the use of torture, so that some utilitarians may consider torture to be morally wrong in these cases.

It might be argued that torture is morally justified only in exceptional cases. One of them might be the following case. A group of criminals have kidnapped a child and threaten to kill it. The police apprehend an accomplice who knows where the child is being hidden, but does not wish to convey this information to the police. The police is incapable of using any non-coercive or coercive strategy in an effective manner to persuade the accomplice to reveal where the child is being kept. In order to save the child, torturing the accomplice proves to be the only option. The police torture her, she says where the child is and the police saves the life of the child. Although torture is illegal in most countries, including all European countries, it turns out to be the most moral action that could have been applied in this specific case.

Returning to torture as a paraphilia, is it an immoral practice? If consenting adults engage in it because they derive pleasure from practicing it, there is not an obvious reason why it should be considered as immoral. The only argument against it might come from virtue ethicists who could claim that torturing a fellow human (or animal) transforms us into less admirable people. Moreover, by torturing we can develop aggressive and sadistic proclivities which at a certain point we will not be able to control and consequently may begin to practice them with non-consenting humans. This would make this paraphilia immoral.

Violence

> Who meditates oppression, his dwelling is overturned. (Babylonian. Hymn to Samas. ERE V. 445)

Rape and torture are only two types of violence among many more. In general, violence (from a realistic and not an idealistic perspective[12]) can be morally justified under certain conditions. The main conditions are that it is reactive and that it does not exceed the violence or threat of violence to which it as a reaction. "Reactive violence" denotes those types of violence that are a response to other violence or threats of violence.

In most cases physical violence should not be a response to non-physical violence. According to such reasoning, violence of a sort may be responded to by violence of the same sort (and should not exceed it). According to a different reasoning, violence of some sort can be rightfully responded to by violence of any other sort, provided that it is not excessive. For example, a mild physical response to verbal sexual harassment might be an understandable and justified reaction.

But what about violence as a paraphilia? As torture is a form of (severe) violence, whatever has been argued about torture as a paraphilia in the previous paragraphs, applies by analogy to violence. If it is consensual, it is not morally wrong—with the reservations previously mentioned regarding torture as a paraphilia.

Sadism

Sadists may be less likely than other sex offenders to show cognitive distortions that justify moral transgressions, since an understanding of the immorality of their actions (causing harm) is precisely what facilitates sexual gratification. Thus, unlike psychopaths, they appear to understand the wrongness of their actions. Furthermore, unlike psychopaths, they are not merely indifferent to the suffering of others. They enjoy it.[13]

---

[12] The "idealist" perspective is that violence can never be justified. According to "idealists", morality requires humans to make an attempt to acquire perfection. Perfect, ideal behavior excludes violence. Whether such behavior is realistic is beyond the point. The "idealist" position is pacifist, according to which violence is always immoral. The utilitarian position argues that violence can be morally justified if it brings about a net "good" for society.

[13] For a recent study suggesting that psychopaths may show some reward-based responses to causing harm, see Decety et al. (2009).

Zoophilia and Bestiality

Zoophilia refers to a human being sexually aroused by an animal. Bestiality refers to humans actually having sex with animals. As it is frequently not possible to know for certain what animals think and feel, animals cannot offer consent. Hence, humans cannot have consensual sex with animals. It is therefore immoral for humans to have sex with animals.

Paedeophilia

> Great reverence is owed to a child. (Roman. Juvenal, xiv. 47)
>
> Children, old men, the poor, and the sick, should be considered as the lords of the atmosphere. (Hindu. Janet, i. 8)
>
> The Master said. Respect the young. (Ancient Chinese. Analects, ix. 22)
>
> The killing of the women and more especially of the young boys and girls who are to go to make up the future strength of the people, is the saddest part… and we feel it very sorely. (Redskin. Account of the Battle of Wounded Knee. ERE V. 432)

Children under a certain age cannot offer *informed* consent. Hence, having sex with children under this age is immoral. Paedophilia is a clear case of a paraphilic moral transgression. But what if the paraphilia is limited to the paedophile phantasizing about sex with children, without even thinking of bringing her phantasies into practice? She does not want to realize her paraphilia because she considers sex with children morally abhorrent. This type of paedophile does not behave immorally. It is even laudable that she fully abstains from all sex in her life, if the only sex she could have pleasure in is sex with a child. Nonetheless, her paedophilia remains a sexual deviancy.

In conclusion, paraphilias can be both sexually deviant and immoral (e.g., sex with infants). But they also can only be sexually deviant without being immoral (mechanophilia) or, conversely, immoral without being clearly *sexually* deviant (instances of rape). There are various ways how paraphilias can be addressed. In many cases MBE could be useful as it could motivate certain types of immoral paraphilics to abstain from their desired activities. Public deliberation with a lot of medical expertise involved in it could help the population at large better understand what is wrong and what is not wrong in certain paraphilias. Moral education can help in those cases in which there is a public consensus about the immorality of a certain paraphilia, while a number of people suffering from this paraphilia don't understand the moral wrongness of it. Exhibitionists and frotteurists may be examples of people who in certain cases lack such an understanding. In those cases moral education can be useful. In cases in which they do understand the moral wrongness of their paraphilia, MBE may be of help.

7.4 The Evil Minding of Others' Business: Reality TV as a Platform for Mass Gossip that Is Morally Inferior to Pornography

> Utter not a word by which anyone could be wounded. (Hindu. Janet, p. 7)
>
> Thou shalt not hate thy brother in thy heart. (Ancient Jewish. Leviticus 19:17)
>
> Love thy neighbour as thyself (Ancient Jewish. Leviticus 19:18)
>
> Love the stranger as thyself (Ancient Jewish. Ibid. 33, 34)
>
> Do to men what you wish men to do to you. (Christian. Matthew 7: 12)
>
> I am a man: nothing human is alien to me. (Roman. Terence, Heaut. Tim.)
>
> He whose heart is in the smallest degree set upon goodness will dislike no one. (Ancient Chinese. Analects, iv. 4)
>
> I have not caused hunger. I have not caused weeping. (Ancient Egyptian. EREy. 478)
>
> Never do to others what you would not like them to do to you. (Ancient Chinese. Analects of Confucius, trans. A. Waley, xv. 23j cf. xii. 2)
>
> Has he... driven an honest man from his family? broken up a well cemented clan? (Babylonian. List of Sins from incantation tablets. ERE v. 446)
>
> Nature urges that a man should wish human society to exist and should wish to enter it. (Roman. Cicero, De Ojficiis, i. iv)
>
> By the fundamental Law of Nature Man [is] to be preserved as much as possible. (Locke, Treatises of Civil Govt. ii. 3)
>
> Speak kindness... show good will. (Babylonian. Hymn to Samas. ERE v. 445)
>
> When the people have multiphed, what next should be done for them? The Master said. Enrich them. Jan Ch'iu said. When one has enriched them, what next should be done for them? The Master said. Instruct them. (Ancient Chinese. Analects, xiii. 9)
>
> Men were brought into existence for the sake of men that they might do one another good. (Roman. Cicero. De Off. i. vii)
>
> Man is man's delight. (Old Norse. Hdvamdl 47)

If Reality TV is designed to cause humiliation and suffering in others for the enjoyment of the audience and profit for the producers, Reality TV is immoral. Even if someone is ready to be publicly humiliated in exchange for money, the practice remains immoral. But what about pornography? While prostitution is not public, pornography is. The question is whether porn stars are publicly humiliated. It does not seem to be the case, as they are stars. On the other hand, some Reality TV participants also become stars. It is therefore difficult to argue on this basis that Reality TV is immoral.

Hence, the reasons for the immorality of Reality TV has not to be sought in the behavior of Reality TV participants, but in the minds of Reality TV viewers? Why do they watch such programs? If one finds that she is entertained by the suffering and humiliation of others, there appears to be a moral problem in her mind. While an occasional instance might be innocent, a regular schedule of such pleasure is another matter. That is also a difference with much of pornography. Most people watch pornography in order to become sexually aroused; the minority watch it in order to see women or anyone else being humiliated. On the other hand, in Reality TV humiliation and suffering is frequently the reason why some people watch it regularly. In that sense, Reality TV is generally immoral, whereas pornography frequently is not.

The fact that in Reality TV we do not witness events in front of us but rather on TV, probably helps us to perceive it as somewhat fictional. It might be similar to inflicting harm or death to people who one does not face directly, but rather through the snipers of sophisticated bombers or rocket launchers.[14]

Reality TV is a typical example of a morally dubious public activity. There is however not a sufficiently broad consensus among most people in many countries to ban it. Hence, subjecting a large part of the population to MBE would have to be compulsory in this specific case. It has already been shown however why compulsory MBE is to be avoided. It seems therefore that public deliberation leading to moral education in this field could be of help to lower the interest in Reality TV and possibly even create the conditions for completely banning it.[15]

References

Campbell, Jeremy. 2002. *The Liar's Tale: A History of Falsehood*. New York, NY: W.W. Norton.
Decety, J., K.J. Michalska, Y. Akitsuki, and B.B. Lahey. 2009. Atypical Empathic Responses in Adolescents with Aggressive Conduct Disorder: A Functional MRI Investigation. *Biological Psychology* 80 (2): 203–211.
Lewis, C.S. 1943. *The Abolition of Man*. Oxford: Oxford University Press.
Quinsly, Vernon L. 2011. Pragmatic and Darwinian Views of the Paraphilias. *Archives of Sexual Behavior* 41 (1): 217–220.
Rakić V. 2020. "Prostitutes, Sex Surrogates and Sugar Babies". Sexuality and Culture. Online first: https://link.springer.com/article/10.1007/s12119-020-09702-y
Sinnott-Armstrong, Walter, and Franklin G. Miller. 2013. What Makes Killing Wrong? *Journal of Medical Ethics* 39 (1): 3–7.

---

[14] It is of course also possible that the direct visual perception of suffering can have a positive effect on human minds. It may namely lead to humanitarian aid to those who suffer or in certain cases even to armed humanitarian intervention against states that inflict severe harm on their citizens (see: https://www.usip.org/publications/2002/08/ethics-armed-humanitarian-intervention; last accessed on 1 November 2020).

[15] For a number of interesting insights, see https://www.thoughtco.com/ethics-and-reality-tv-4016356; last accessed on 1 November 2020.

Chapter 8
Good 1

8.1 Empathy

> He who is asked for alms should always give. (Hindu. Janet, i. 7)
>
> The poor and the sick should be regarded as lords of the atmosphere. (Hindu. Janet, i. 8)
>
> Whoso makes intercession for the weak, well pleasing is this to Samas. (Babylonian. ERE v. 445)
>
> Has he failed to set a prisoner free? (Babylonian. List of Sins. ERE v. 446)
>
> I have given bread to the hungry, water to the thirsty, clothes to the naked, a ferry boat to the boatless. (Ancient Egyptian. ERE v. 446)
>
> One should never strike a woman; not even with a flower. (Hindu. Janet, i. 8)
>
> There, Thor, you got disgrace, when you beat women. (Old Norse. Hdrbarthsljoth 38)
>
> 'In the Dalebura tribe a woman, a cripple from birth, was carried about by the tribespeople in turn until her death at the age of sixty-six.'... 'They never desert the sick.' (Australian Aborigines. ERE v. 443)
>
> You will see them take care of., widows, orphans, and old men, never reproaching them. (Redskin. ERE v. 439)
>
> Nature confesses that she has given to the human race the tenderest hearts, by giving us the power to weep. This is the best part of us. (Roman. Juvenal, xv. 131)
>
> They said that he had been the mildest and gentlest of the kings of the world. (Anglo-Saxon. Praise of the hero in Beowulf, 3180)
>
> When thou cuttest down thine harvest... and hast forgot a sheaf, thou shalt not go again to fetch it: it shall be for the stranger, for the fatherless, and for the widow. (Ancient Jewish. Deuteronomy 24:19)

> There are two kinds of injustice: the first is found in those who do an injury, the second in those who fail to protect another from injury when they can. (Roman. Cicero, De Off. I. vii)
>
> Men always knew that when force and injury was offered they might be defenders of themselves; they knew that howsoever men may seek their own commodity, yet if this were done with injury unto others it was not to be suffered, but by all men and by all good means to be withstood. (English. Hooker, Laws of Eccl. Polity, I. ix. 4)

Empathy is one of the most crucial moral values of the human. It is the basis of altruism and of the Golden Rule: to do to others that what you would others like to do to you. According to one group of scholars, humans have been born selfish, while altruism developed through evolution as a mechanism that can help in protecting in-groups. According to another group of scholars, humans have been born with a sense of altruism (Bloom 2013). In that case, the question ought to be asked what has led to the development of selfishness in humans.

8.2 Justice, Rertribution

> Justice is the settled and permanent intention of rendering to each man his rights. (Roman. Justinian, Institutions., I. i)
>
> The first point of justice is that none should do any mischief to another unless he has first been attacked by the other's wrongdoing. The second is that a man should treat common property as common property, and private property as his own. There is no such thing as private property by nature, but things have become private either through prior occupation (as when men of old came into empty territory) or by conquest, or law, or agreement, or stipulation, or casting lots. (Roman. Cicero, De Off. I. vii)
>
> Do no unrighteousness in judgement. You must not consider the fact that one party is poor nor the fact that the other is a great man. (Ancient Jewish. Leviticus 19:15)
>
> I have not traduced the slave to him who is set over him. (Ancient Egyptian. Confession of the Righteous Soul. ERE v. 478)
>
> Regard him whom thou knowest like him whom thou knowest not. (Ancient Egyptian. ERE v. 482)

The physical abuse of a helpless child or disabled person understandably provokes among many people a degree of aggressiveness toward the abuser. If this aggressiveness leads to an action that protects the abused individual in an appropriate manner, this action has a positive moral value. Also, if a judge is overwhelmed by empathy toward a murderer who she should sentence to an adequate number of years in prison, the judgment she passes might not be appropriate if her empathy with the accused adversely interferes with the moral value of just retribution. She will act appropriately only if she rules in a just way, applying the most apposite, legally possible retribution to the murderer.

8.3 Evolutionary Morality—Did It Make Us Good or Evil?

> Brothers shall fight and be each others' bane. (Old Norse. Account of the Evil Age before the World's end, Volospd 45)
>
> Has he insulted his elder sister? (Babylonian. List of Sins. ERE v. 446)
>
> To take no notice of a violent attack is to strengthen the heart of the enemy. Vigour is valiant, but cowardice is vile. (Ancient Egyptian. The Pharaoh Senusert III, cit. H. R. Hall, Ancient History of the Near East, p. 161)
>
> They came to the fields of joy, the fresh turf of the Fortunate Woods and the dwellings of the Blessed … here was the company of those who had suffered wounds fighting for their fatherland. (Roman. Virgil, Aeneid, vi. 638–9, 660)
>
> Courage has got to be harder, heart the stouter, spirit the sterner, as our strength weakens. Here lies our lord, cut to pieces, out best man in the dust. If anyone thinks of leaving this battle, he can howl forever. (Anglo-Saxon. Maldon, 312)
>
> Love thy wife studiously. Gladden her heart all thy life long. (Ancient Egyptian. £i^£v. 481)
>
> Nothing can ever change the claims of kinship for a right thinking man. (Anglo-Saxon. Beowulf, 2600)
>
> Is it only the sons of Atreus who love their wives? For every good man, who is right-minded, loves and cherishes his own. (Greek. Homer, Iliad, ix. 340)
>
> The union and fellowship of men will be best preserved if each receives from us the more kindness in proportion as he is more closely connected with us. (Roman. Cicero. De Off. i. xvi)
>
> If any provide not for his own, and specially for those of his own house, he hath denied the faith. (Christian. I Timothy 5:8)
>
> Did not Socrates love his own children, though he did so as a free man and as one not forgetting that the gods have the first claim on our friendship? (Greek, Epictetus, iii. 24)
>
> Natural affection is a thing right and according to Nature. (Greek. Ibid. i. xi)
>
> I ought not to be unfeeling like a statue but should fulfil both my natural and artificial relations, as a worshipper, a son, a brother, a father, and a citizen. (Greek. Ibid., lll.ii)
>
> This first I rede thee: be blameless to thy kindred. Take no vengeance even though they do thee wrong. (Old Norse. Sigdrifumdl, 22)
>
> Part of us is claimed by our country, part by our parents, part by our friends. (Roman. Ibid. i. vii)

If a ruler ... compassed the salvation of the whole state, surely you would call him Good? The Master said. It would no longer be a matter of "Good". He would without doubt be a Divine Sage. (Ancient Chinese. Analects, vi. 28)

Has it escaped you that, in the eyes of gods and good men, your native land deserves from you more honour, worship, and reverence than your mother and father and all your ancestors? That you should give a softer answer to its anger than to a father's anger? That if you cannot persuade it to alter its mind you must obey it in all quietness, whether it binds you or beats you or sends you to a war where you may get wounds or death? (Greek. Plato, Crito, 51, a, b)

'Put them in mind to obey magistrates.'... 'I exhort that prayers be made for kings and all that are in authority.' (Christian. Titus 3:1 and I Timothy 2:1, 2)

Your father is an image of the Lord of Creation, your mother an image of the Earth. For him who fails to honour them, every work of piety is in vain. This is the first duty. (Hindu. Janet, i. 9)

Honour thy Father and thy Mother. (Ancient Jewish. Exodus 20:12)

To care for parents. (Greek. List of duties in Epictetus, in. vii)

Has he despised Father and Mother? (Babylonian. List of Sins. ERE v. 446)

I was a staff by my Father's side ... I went in and out at his command. (Ancient Egyptian. Confession of the Righteous Soul. ERE v. 481)

Rise up before the hoary head and honour the old man. (Ancient Jewish. Leviticus 19:32)

I tended the old man, I gave him my staff. (Ancient Egyptian. ERE v. 481)

You will see them take care ... of old men. (Redskin. Le Jeune, quoted ERE v. 437)

I have not taken away the oblations of the blessed dead. (Ancient Egyptian. Confession of the Righteous Soul. ERE v. 478)

When proper respect towards the dead is shown at the end and continued after they are far away, the moral force (te) of a people has reached its highest point. (Ancient Chinese. Analects, i. 9)

Whence is the population to be kept up? Who will educate them? Who will be Director of Adolescents? Who will be Director of Physical Training? What will be taught? (Greek. Ibid.)

'Nature produces a special love of offspring' and 'To live according to Nature is the supreme good.' (Roman. Cicero, De Off. i. iv, and De Legibus, i. xxi)

Children, the old, the poor, etc. should be considered as lords of the atmosphere. (Hindu. Janet, i. 8)

To marry and to beget children. (Greek. List of duties. Epictetus, in. vii)

The second of these achievements is no less glorious than the first; for while the first did good on one occasion, the second will continue to benefit the state for ever. (Roman. Cicero. De Off. i. xxii)

Praise and imitate that man to whom, while life is pleasing, death is not grievous. (Stoic. Seneca, Ep. liv)

> The Master said. Love learning and if attacked be ready to die for the Good Way. (Ancient Chinese. Analects, viii. 13)
>
> It is upon the trunk that a gentleman works. When that is firmly set up, the Way grows. And surely proper behaviour to parents and elder brothers is the trunk of goodness. (Ancient Chinese. Analects, i. 2)
>
> You will see them take care of their kindred [and] the children of their friends ... never reproaching them in the least. (Redskin. Le Jeune, quoted ERE v. 437)

One of the explanations for the emergence of morality is the preservation of the species. Persson and Savulescu ground their explanation of the development of morality precisely on that. As we can see from the above citations, their explanation has a foundation in the ways of thinking of our ancestors.

But again, the question comes up—have humans been born selfish or altruistic? In the first case, altruism developed in order to strengthen the species' inclusive fitness. In the second case, it remains unclear why humans who were born with a sense if altruism became selfish.

8.4 Fortitude and Politics

> What good man regards any misfortune as no concern of his? (Roman. Juvenal xv. 140)
>
> Death is to be chosen before slavery and base deeds. (Roman. Cicero, De Off. i, xxiii)
>
> Death is better for every man than life with shame. (Anglo-Saxon. Beowulf, 2890)
>
> Nature and Reason command that nothing uncomely, nothing effeminate, nothing lascivious be done or thought. (Roman. Cicero, De Off. i, iv)
>
> We must not listen to those who advise us "being men to think human thoughts, and being mortal to think mortal thoughts," but must put on immortality as much as is possible and strain every nerve to live according to that best part of us, which, being small in bulk, yet much more in its power and honour surpasses all else. (Ancient Greek. Aristotle, Eth. Nic. 1177B)
>
> The soul then ought to conduct the body, and the spirit of our minds the soul. This is therefore the first Law, whereby the highest power of the mind requireth obedience at the hands of all the rest. (Hooker, op. cit. i. viii. 6)
>
> Let him not desire to die, let him not desire to live, let him wait for his time ... let him patiently bear hard words, entirely abstaining from bodily pleasures. (Ancient Indian. Laws of Manu. ERE ii. 98)

> He who is unmoved, who has restrained his senses ... is said to be devoted. As a flame in a windless place that flickers not, so is the devoted. (Ancient Indian. Bhagavad gita. ERE ii 90)
>
> Is not the love of Wisdom a practice of death? (Ancient Greek. Plato, Phadeo, 81 A)
>
> I know that I hung on the gallows for nine nights, wounded with the spear as a sacrifice to Odin, myself offered to Myself. (Old Norse. Hdvamdl, I. 10 in Corpus Poeticum Boreale; stanza 139 in Hildebrand's Lieder der Alteren Edda. 1922)
>
> Verily, verily I say to you unless a grain of wheat falls into the earth and dies, it remains alone, but if it dies it bears much fruit. He who loves his life loses it. (Christian. John 12:24, 25)

Fortitude is frequently associated with a warrior's disposition, and sometimes with a politician's proclivity. The citations in this section show this as well. But what is the relationship between fortitude, politics and ethics now? Has something changed? Where do politicians stand in moral terms and how can they be helped by ethicists? The following section will address this issue, arguing that an important new variable is entering ("new" in comparison with the time of most of the citations), and should continue entering, the relationship between fortitude in politics and ethics. This variable is science.

8.5 Ethics and Politics—Science Might Make Politics More Moral

For a very long time already, not only philosophers but much of the population at large have had an interest in the (im)morality of politicians. Philosophers and other scholars have been reflecting for centuries on the theme how to make politics more moral. I argue that science can give a contribution to ethics permeating the political realm. Scientific evidence can be an important criterion for assessing the morality of politics. The relationship between politics and ethics would then, in part, be mediated by science. Hence, scientific standards can be utilized by ethicists, as well as by a larger educated public engaged in democratic deliberation, in order to pass judgments on morally relevant issues in politics.

The relationship between ethics and politics has been discussed for centuries by numerous philosophers and other scholars. It is also a recurrent theme in public discourse in and about politics. The values of political ideologies, specific policies or political strategies of various kinds, are sometimes assessed by how ethical they are. But we don't always have identical views on what is ethical and what not, or on what is more and what is less ethical.

We even don't have a philosophical consensus on the issue of whether politics ought to be founded on ethical principles (Rawls 1973; Nozick 1974, and many others) or ethics on politics (e.g., Barber (1989) in a strongly formulated version of this argument). Should ethicists (philosophers) determine morality in politics or should this be determined by democratic deliberation?

I propose scientific evidence as an important criterion for assessing the morality of politics.[1] It would not be the only criterion, but one of the criteria by which the ethics of politics is to be assessed. If a policy contradicts scientific evidence or if it is not based on scientific evidence, it is less reliable than a policy that is based on scientific evidence. Political campaigns that have no groundwork in evidence, but in biases or even lies, are not only epistemologically, but also morally inferior to those that do have such a groundwork. Political actors campaigning on political platforms and policies that or not based on assumptions that are true, also discredit the morality of their campaigns.

Hence, in reflecting on the relationship between politics and ethics, scientific evidence can be of significant help in certain cases. In such cases scientific evidence would not only be able to assess the epistemic value of policies or political campaigns, but it would also have implications for their moral value. Science would then mediate between politics and ethics, helping ethics in its assessment of the morality of certain policies, political strategies or campaigns. Various examples can be given. As a clarification, one will follow.

Politician x, running for the office of president of democratic country A, predicts that in this country the current influx of migrants will lead to the numerically dominant nation of country A becoming a national minority in 12 years. This projection is based on the current rates of migrants entering country A and the very high birth rate of one group of migrants. Politician x bases his campaign on this projection and induces fear and xenophobia among the members of the numerically dominant nation in country A. Politician x wins the elections, largely because of the fear of migrants he induced among the population of his country.

Scientific evidence demonstrates however that the projections of politician x were wrong. First, all available scientifically valid indicators showed that the current rate of migrants entering country A cannot last 12 years but for a period of six months at most. After this period, the number of migrants entering country A is expected to steadily decrease and become in less than one year only 10% of the current rate. Second, the birth rate of the ethnic group politician x focused on was indeed unusually high, but this group consisted of only 4% of the number of migrants entering country A. Politician x extrapolated his projections as if the birth rate of these 4% was the birth rate of all migrants entering A. Hence, the projection that the numerically dominant nation of country A would become a minority in 12 years was woefully misplaced.

Nevertheless, politician x won the presidential elections, largely due to assumptions that were not based on scientific evidence. The wrongness of those assumptions was something most voters were not aware of, while the political opponents of politician x were not able to capitalize on their wrongness, because they focused mostly on their own campaigns and the differences among them, at the same time casting

---

[1] I take scientific evidence to consist of scientifically testable data indicating whether an assumption is true.

doubt on politician x because he was not the political mainstream and *not* because his campaign was based on misinterpreted evidence and wrong projections. Had they focused on this, politician x might have lost the elections. In that case, scientific evidence would have had a moral value because of the following likely outcomes:

1. The population would not have been misinformed.
2. The population would not have been distressed by unfounded fears and, as a result, xenophobic reactions of various types would have been less frequent and strong in country A.
3. The presidential elections might have been won by a politician who had not founded his campaign on wrong assumptions, xenophobic biases and fear; hence, it is reasonable to assume that those political actors would have been victorious who had acted in a more ethical manner than politician x had; they would have deserved to win the elections because their campaigns would have been more evidence-based and more ethical.
4. Had the presidential elections been won by someone whose campaign had not been based on wrong assumptions but on evidence, the likelihood would have been higher that an ethically more appropriate policy would have been conducted in country A in the years following the elections.

Belief in biases, neglect of scientific or ethical expertise and dismissal of evidence are problems that affect politicians in various political systems. Democracies are certainly no exception: political systems in which people freely elect their representatives face the danger of becoming platforms for contests of non-evidence based opinions. Democratic deliberation is also not immune to this danger.

Although not all ethical issues that politics gives rise to can be addressed by scientific evidence, my argument is that some of them can. If a political position can be shown to be founded on something that is disputed by science, this position is not based on true premises. If a political campaign is founded on something that is not based on true premises, it is a campaign that is not conducted on the basis of proper ethical standards. *Hence, science can be a standard used by ethicists to pass judgments on morality in politics.*

Moreover, if scientists in a country take an active role in publicly assessing certain policies on the basis of scientific evidence, the policies in that country may become better and therefore more in the interest of the population. Consequently, they would become more moral. Politics and ethics, mediated by science, would then establish a closer and more useful relationship in the political realm. Ethics would utilize an important instrument for passing judgment on moral issues permeating this realm. This instrument would be scientific evidence.

The population in many countries is more inclined to believe scientists than politicians. This would make the work of scientists easier. They could benefit their countries, at the same time not becoming involved in politics as politicians but as those who determine to what extent political strategies, policies and campaigns are based on scientific evidence and consequently on truth and proper moral standards.

Furthermore, the population may become gradually educated/more aware of the importance of respecting scientific evidence. It will therefore not always be necessary that a scientist appears in mass media and explains what is evidence based and what not. An educated general population might in certain cases judge for themselves.

The precondition for this is that the public is aware of the relevance of scientific evidence and that it acts in line with this awareness, instead of succumbing to political, ideological and other biases, or to political manipulation. An essential step in order to achieve this is public education leading to the development of a political culture that is very sensitive to both the epistemic and the moral value of scientific evidence in the political sphere. A politician's fortitude would consist in his public stance supporting the importance of scientific evidence and his active support for programs explaining the electorate why it is important to pay attention to this evidence, rather than to political campaigns and short-term political gains by politicians.

References

Barber, B. 1989. *The Conquest of Politics*. Princeton: Princeton University Press.
Bloom, P. 2013. *Just Babies: The Origins of Good and Evil*. New York, NY: Crown Publishers.
Nozick, R. 1974. *Anarchy, State and Utopia*. New York: Basic Books.
Rakić, V. 2017. Compulsory Administration of Oxytocin Does Not Result in Genuine Moral Enhancement. *Medicine, Health Care and Philosophy* 20 (3): 291–297.
Rawls, J. 1973. *A Theory of Justice*. Oxford: Oxford University Press.

Chapter 9
Good 2

9.1 Making Moral Enhancement Work

In the preceding arguments it has been demonstrated that both moral education and MBE can help humans become better. It has also been argued that becoming better entails an essential added value: those who become better tend to become happier. The critical question is how to bring moral education and MBE into practice. In what follows it will be discussed how to bring moral enhancement into practice under the circumstances we deal with.

First of all, as I argued in the preceding section, political decision makers should be taught to value scientific evidence. If it is scientifically proven that morality and happiness are in a circularly supportive relationship and that moral behavior induces happiness in most people most of the time, moral enhancement can be a program politicians have reason to campaign for. That means that scientists and ethicists can conduct their own campaign aimed at raising awareness among politicians about the benefits of moral enhancement. The initiative has to come from scientists and ethicists. Although "campaigning" is not an activity these people are generally associated with, the enhancement of morality of humans is an objective they have reason to consider worthy enough to raise awareness about it among the population at large.

In that case, the moral values reflected in the citations showing the trans-cultural and trans-geographical similarities between various civilizations can be enhanced. Scientific evidence, cognitive enhancement, moral education and MBE can help in that. They can contribute to enhancing the moral evolution of humans from the status being reflected in some of the citations from *the Abolition of Man*, to a new morality, an Ultimate Morality.

Societies that have a tradition of respecting scientific findings will have an important advantage—an easier job in motivating politicians to act ethically, on the basis of the latest findings and theories on moral enhancement. Those societies will have a leading role in adopting moral enhancement policies. Their morally enhanced populations can be expected to be happier than those populations that, ignoring scientific evidence, will consider moral enhancement, especially MBE, as superfluous or harmful activities.

The scientifically and ethically more aware societies will have policy makers who are more likely to adopt affirmative action policies for the morally enhanced, including those who have successfully completed certain moral education programs (programs that governments can fund) or who have subjected themselves to MBE. It might even become a matter of prestige for a country to have effective moral enhancement programs. That would not only suggest that such a country cares about morality, but also that it cares about scientific evidence.

Does this approach contain the danger of societies becoming totalitarian, to a certain extent? If states provide "advantage of opportunity" to the morally enhanced by monitoring how much moral education they have been through and whether they have subjected themselves to MBE, isn't such a state excessively intruding into the private spheres of its citizens?

It does not seem so. Our private sphere has been intruded in a variety of ways already (e.g., acquisition of genetic data or health data in general). Moral enhancement programs, on the other hand, wouldn't necessarily acquire our private data. Moreover, such programs could be developed by the most competent ethicists. Additionally, these programs would be voluntary. The advantage of opportunity strategy is not coercion. It is rather nudging. Finally, governments don't *have* to adopt affirmative action policies. Awareness raising about the benefits of moral enhancement (including MBE) will do most of the work. Affirmative action policies favoring the morally enhanced would only be an additional stimulus for people to enter programs of moral enhancement.

Those countries that take moral enhancement seriously will also have the opportunity to cooperate in that field. In particular, if developed and influential states decide to cooperate in the field of moral enhancement, it is possible to envision them signing protocols to that effect. It is easy to imagine that, if moral enhancement starts to be taken seriously by governments, protocols on moral enhancement, similar to the Kyoto protocol on environmental issues, could be signed and implemented.

As the likelihood is high that developed countries will be the first to implement moral enhancement policies, it is imaginable that groups of countries similar to the G7 will develop common policies on moral enhancement. Or, if the EU Parliament or Commission adopt such policies, EU member states might do the same. The United Nations may also adopt policies to the same effect. All in all, moral enhancement policies are not a matter of imagination. They are rather a matter of realistically perceiving the world around us.

9.2 A Morally Superior Posthuman Species Guided by Ultimate Morality

When did the future switch from being a promise to being a threat?

Chuck Palahniuk

Ethics is oriented toward the future. Any activity that aspires to transcend the present is future-oriented. So is the aspiration of ethicists to deliberate on how humans can become better.

As all futuristic arguments, this chapter contains speculative elements. Moreover, the conception of UM that will be developed might be repulsive to some readers, because it accepts, in a very specific form, a transformative change of the human species as the most moral outcome of the future. Hence, those who dislike futuristic speculations and who believe that survival of the human species (as it is) is a moral must, might consider concluding the book at this point. They will however miss its essential point.

This chapter will primarily show how a posthuman future in which the comprehension-motivation gap has been surpassed by voluntary MBE (VMBE) could look like. In addition, it will offer a deductive argument why the human species ought not to aspire to survive at any cost, but rather accept UM as a promising moral philosophy. What follows are mostly succinct logical reflections on the future. They are of essential importance for this book.

9.3 Conceptual Clarifications

The following conceptual definitions are relevant for the arguments that will follow:

- *"Posthumans" are beings with a higher moral status than "mere humans".*
- *"Mere humans" are members of the currently existing human species.*
- *"Higher moral status" can imply not only cognitive superiority, but superior moral dispositions as well, that is, a higher moral value of an agent's actions or character; moral superiority is precisely the criterion by which I will define beings with a higher moral status, that is, posthumans; hence:*
- *Posthumans are beings that are morally superior to "mere humans".*

Moral enhancement includes not only the enhancement of someone's cognitive abilities to understand morality, but also encompasses an augmented inclination to act in accordance with those abilities. It entails superseding the comprehension-motivation gap. Hence, it will be argued that our motivation to *act* morally (that is, in line with how we believe we *ought* to act) is an essential disposition of beings with a superior moral status, that is, of posthumans (see also Rakić 2015).

9.4 Ultimate Morality and the Human Emancipation from the Survival-at-Any-Cost Bias

The following two statements are true[1]:

(1) the creation of posthumans is imaginable, if they are envisaged as morally enhanced beings who have transcended the comprehension-motivation gap;
(2) the creation of posthumans is a moral duty, subject to the condition that we create morally enhanced posthumans, as defined in (1).[2]

The truthfulness of statement (1) is self-evident. The truthfulness of statement (2) is based on two arguments. The first is inductive, indicating probability. The second is deductive, offering proof.

9.4.1 Inductive Argument—Probability

In an *inductive* argument, the truth of all its premises is logically compatible with the falsehood of its conclusion. It is therefore not a proof, but it does indicate probability.

If we assume that the higher moral status of posthumans implies their enhanced morality, we may argue, inductively, that (morally enhanced) posthumans will not be inclined to annihilate mere humans. For if mere humans have moral inhibitions against obliterating some species of a lower moral status than their own, morally enhanced posthumans will be even less likely to do the same to mere humans. In fact, they might consider it their moral duty to preserve those beings who enabled them to come into existence. Hence, our duty to enhance morality, that is, to create or help the eventuation of morally enhanced posthumans is not against our interests. It is a duty mere humans can realize.

The fact that posthumans will not be inclined to annihilate or harm mere humans, does of course not mean that they will refrain from it under all circumstances. The issue of whether posthumans choose to preserve mere humans depends on the opportunity costs of doing so. Differences in moral status are supposed to help us to correctly make moral trade-offs. The difference in moral status between humans and sheep suggests that it would be right to inflict suffering on the latter if doing so was the only way to get a vaccine for COVID-19. Hence, it is possible that posthumans judge their objectives to be very morally important and assess as necessarily requiring painful experiments on mere humans with inferior moral status. For an argument along these lines, see Agar (2013).[3]

---

[1] The following paragraphs build on Rakić (2021).

[2] Mere humans "creating" posthumans should be understood rather broadly. It includes mere humans *helping* the eventuation of posthumans.

[3] I am thankful to an anonymous reviewer who formulated this argument.

9.4.2 Deductive Argument—Proof

Unlike inductive arguments, deductive arguments count as proofs in that their conclusions are contained in the premises. The deductive argument in favour of the creation of posthumans is the following: even if morally enhanced posthumans should decide to annihilate mere humans, such a decision is by necessity a morally superior stance to the wish of mere humans (that is, morally unenhanced humans) to continue to exist. This is a deduction, following from the following two true premises:

1. Morally enhanced humans make better moral judgment than mere humans.
2. One of the attributes of posthumans, as we have defined them, is that they are morally enhanced.

From these two true premises follows a conclusion that is also true:

3. Posthumans make better moral judgments than mere humans.

I will call this type of moral reasoning and behaviour Ultimate Morality. Here is an expanded version of its definition:

- *Ultimate Morality refers to a morally enhanced reasoning and behaviour, marked by a subjugation of personal interests and the interests of our species to those stances that we consider as morally superior. This might include the decision of moral agents to accept harm for themselves, and even to be sacrificed for the sake of morally superior posthumans. Hence,* **UM is the supreme moral foundation of the human emancipation from the survival-at-any-cost bias**.[4]

This emancipation has an additional feature. Even if we practice UM as an ethics that will result in the creation of a species other than the human species, we will not always know in which specific instances the other posthuman species is morally superior to the human species. We will only know that it is by definition so, as we have defined posthumans as being morally superior to mere humans. Our understanding of this is similar to that of the theodicy, the moral justification of which we have difficulty to comprehend. This time it is however not God but posthumans who practice moral superiority that is not fully understandable to humans. Humans may in this way better comprehend the theodicy. It will become more comprehensible to them that some events that are morally acceptable are not understandable to humans as being moral. Humans may better comprehend that their paths are different from God's or posthuman paths.

Clearly, it would be preposterous to expect from existing mere humans to sacrifice their lives for morally superior posthuman individuals. But mere human individuals do not have to be expected to do that. First, the inductive argument indicates that morally superior posthumans are unlikely to sacrifice those who made them come into being (that is, mere humans). Hence, posthumans can be created de novo, without harming existing humans. Second, mere humans, guided by UM, can *gradually*

---

[4]See Rakić (2014).

evolve into a morally superior posthuman species. Such a gradual evolvement would not harm existing mere humans, and probably not their close descendents. On the contrary, there are all reasons to believe that it would benefit them.

The described type of morality is called "ultimate", because of two reasons. First, it is a morality that is by definition superior, as its objective is a perfectly moral world. Second, it is "ultimate" because its aim of a morally perfect world is the final purpose of morality. *Moreover, UM imposes a duty on mere humans that consists of mere humans helping the eventuation of morally enhanced posthumans. It has been previously shown that moral enhancement implies an increase in the net balance of happiness. UM imposes a **duty** on mere humans that eventually will result in an increase of the net balance of happiness in the world.*[5] ***In that sense, UM can even be a basis for reconciling deontology and utilitarianism.***

The survival-at-any-cost contention is a reasoning that remains within the confines of the "ultimate harm" argument. But it is not a morally justified argument, as has been shown in the above deduction. In moral terms, UM trumps ethical theories grounded in "ultimate harm" prevention. These theories suffer from a biology-based survival imperative.

On the other hand, UM is an ethical theory that is free from biological underpinnings based on the survival-at-any-cost bias. It is an autonomous ethical theory that does not have a groundwork in anything other than "pure" ethics, that is, the moral duty to help the eventuation of a species of posthumans that is defined as morally superior to existing humans. Practicing UM, humans will act in line with their duty and in line with their happiness maximization.

9.5 Posthuman Morality: Morality Ultimate, Comprehension-Motivation Gap Superseded

Beings who always behave in line with what they believe to be moral have been defined here as beings with a higher moral status than mere humans have. This is so because the difference between beings who are capable of moral reasoning only and those who practice their moral beliefs is a qualitative difference amounting to a differentiation in moral status. The morality of the latter pertains to beings with a higher moral status. It is posthuman morality. Moral enhancement, if understood as an intervention not just making us comprehend morality better but also resulting in us *behaving* more morally, is an intervention with such massive implications that it amounts to nothing less than a moral status enhancement, and consequently to the creation of posthumans (see Rakić 2015).

What has been said about posthumans here pertains to a morality the center of which is the individual. Those who have transcended the comprehension-motivation gap are those individuals who always behave as they believe they ought to behave.

---

[5]This net balance should sometimes be measured on the basis of the value of median happiness, as has been shown in Part I of this book.

They are always good. Posthuman morality is marked by behaviour that is *always* morally apposite. That is how we defined posthumans. Posthumans treat others *always* as they would like to be treated by others. But posthuman morality is not merely directed to the benefit of individuals. It is also directed to the moral benefit of the world. *Posthumans are the ones who will occupy the empty set from Part I, Chap. 5.2.*

There are two scenarios of UM that mere humans can follow:

1. To help facilitate a *gradual* development of posthumans from the human species (e.g., after the third or fourth generation) or to help the eventuation of posthumans de novo. In this scenario existing humans could be spared harm. The human duty to help the eventuation of posthumans would be compatible with the interests and happiness of existing humans. This is the scenario that reconciles duty and happiness, bringing together deontology and utilitarianism.
2. To help the facilitation of an abrupt replacement of existing mere humans with posthumans. This would imply the annihilation of existing humans and would therefore be incompatible with their interests and happiness. Utilitarian ethics would be divorced from deontology.

In both scenarios mere humans become posthumans, by virtue of adopting UM. The only scenario that is however acceptable to existing mere humans is the first one, as in that scenario they do not have to be harmed.

Posthumans are not beings who return to the status-quo-ante vis-a-vis the Original Sin. They don't return to the Garden of Eden. Their status is not the human status as described in the Garden of Eden. It differs from it in that posthumans know what evil is. In the Garden of Eden there was no knowledge of evil. After the Original Sin thinking good ceased to imply acting good. The Original Sin has created the comprehension-motivation gap. Unlike humans (before and after the Original Sin[6]), posthumans know evil, but don't act evil. They have superseded the comprehension-motivation gap.

9.6 Kant

When we talk about the duty of humans to devote themselves to moral enhancement, as well as the moral duty to accept UM, even if that includes a preference for the existence of posthumans to mere humans, the question comes up how to explain the possible development of such a preference. An answer might be found in the Kantian concept of "ethical maturation" and "moral progress" of humankind.

---

[6]Before the Original Sin humans didn't know of Evil, while after the Original Sin they know of it but sometimes act precisely in line with it.

VMBE can speed up the process of ethical maturation/moral progress of humanity. If VMBE can be instrumental in helping humans surpass the comprehension-motivation gap, it will help them in their ethical maturation. It will help them in their moral progress.

Such a moral progress does however not include only an enhanced motivation to act more morally. It encompasses also enhanced moral reflection. An essential feature of such an enhancement is a lower level of heterophobia. This means that a human with enhanced moral reflection will be more inclined to ascribe equal moral status to racially, sexually, confessionally, ethnically and nationally different people. A cosmopolitan worldview is therefore to be expected in morally enhanced humans—in an ethically matured humanity. Such a worldview also implies cosmopolitan thinking in politics. In its final instance, it implies the establishment of an ethical world state.

Such a state has resemblances to Kant's "Ethical Commonwealth", understood as a community guided by virtue alone (Kant 1907). This community is therefore different from a world state that is guided by political interests. It is an ethical community. Hence, in Kant's world state the cleavage between ethics and politics is superseded. The "Ethical Commonwealth" is a world state, but a world state that is perfect in an ethical sense. It is a community in which the gap between ethics and politics has been superseded, in the sense that politics has become part of ethics. In Kant's "Ethical Commonwealth" the gap between the "is" (politics") and the "ought" (ethics) has been surpassed in favour of the "ought". Politics has been absorbed by ethics. The concept of the "Ethical Commonwealth", moreover, is not confined to specific states but extends to humanity in general (Kant 1907, Ak. 6:96).

9.7 A World State of the Future—Again

*Hail **Earth** mother of us all.*

Pagan goddess; The Rites of Odin

Let my country die for me.

James Joyce; Ulysses

Human augmented ethical maturity in a futuristic version of Kant's Ethical Commonwealth implies, among else, the following features: 1. VMBE; 2. consequently, more happiness (if posthumans eventuate gradually from mere humans or if they are created de novo); 3. UM. *In feature 2 ulitarianism would be reconciled with deontology. Deontological practice would result in a utilitarian outcome. By fulfilling our moral duties based on UM we would gradually create a world inhabited by posthumans who approach moral and thus felicific perfection. Hence, it is possible to reasonably expect a futuristic version of Kant's Ethical Commonwealth/global state, that is guided by a morality in which deontological and utilitarian principles are squared.*

9.7 A World State of the Future—Again

Although ever more efficient and safe MBE technologies are likely to be developed in the time to come, serious challenges for moral enhancement remain:

- moral reflection cannot be enhanced by MBE technologies, and it is precisely moral reflection that is needed in morally complex contexts;
- as the only true type of MBE (for existing individuals capable of decision making) is VMBE, the question is what will motivate people to voluntarily embark on the MBE enterprise.

The first challenge can be addressed by traditional moral enhancement (moral education). It has also been demonstrated that being good is generally conducive to happiness. Hence, humans will become more motivated to act morally with the help of MBE. Our moral reflection might not improve as much as would be useful, but our moral motivation would obtain a significant boost: we would come ever closer to bridging the comprehension-motivation gap. We would gradually return to the Garden of Eden, but this time with knowledge of Evil. Knowing Evil, however, would not make us act in an evil way. *The reason is that primarily VMBE is likely to bring us ever closer to superseding the comprehension-motivation gap by enhancing our motivation to be good.*

UM is unimaginable in an order that is confined to specific communities. UM is universal. In line with the contention that Kant's vision of the (not immediate but more distant) future of humanity is one of a cosmopolitan moral order in which humans act virtuously in the broadest possible community, that is, humanity (Kant 1907), it is justified to conclude that successful VMBE is conducive to Kant's position. It is conducive for the following two reasons:

1. successful VMBE will make it likelier for humans to behave virtuously not in a confined community to which their morality extends, but in the broadest possible human community (the "Ethical Commonwealth"); and
2. successful VMBE, augmenting our inclination to act morally, will shorten the time needed for humans to establish such a community.

The Ethical Commonwealth is not far away if humans are rational. In order to bring it about what is needed is two types of education:

1. Awareness raising about the relatedness of morality to happiness.
2. Awareness raising about the usefulness of MBE for happiness.

Those humans who opt for MBE will better understand the relationship between morality and happiness. The more they use MBE, the more they will find out that moral behaviour increases their happiness—that it tends to be in their interest. It will be rather easy to understand this, especially for educated people, because it is not a speculation—it is founded in science.

After having developed the capacity of causing "ultimate harm", humanity might need UM in order to realize a very specific form of "ultimate harm"—the gradual replacement of humans by posthumans or an abrupt sacrifice of humans in favour of posthumans (the former obviously to be preferred to the latter). The realization of this form of "ultimate harm" might be the ultimate moral purpose of humanity to surpass itself by creating posthumanity. As has been shown, in a gradual replacement of mere humans with posthumans, mere human individuals would not be sacrificed, while deontology and utilitarianism would be squared; in an abrupt replacement there would be no reconciliation of the two.

Man's final conquest has proved to be the abolition of Man, C. S. Lewis

C. S. Lewis is distressed by the "abolition of man" (the annihilation of the human species). The abolition of the human does however not necessarily have to imply something abhorrent. Much to the contrary. UM "abolishes" the human, but it performs the only morally right abolition of the human—replacing the human by a morally superior posthuman.

What are the dispositions of this morally superior posthuman? One of them is freedom, her determination not to sell her soul for power. The other is that she does not aspire the survival of the human species at any cost.

The latter disposition is related to the former. A posthuman species consists of humans who wish to survive as individuals, but not necessarily as a species. If survival as a species brings into question human freedom (e.g., by compulsory MBE) the rationale of humans surviving as a species is being brought into question.

It is precisely freedom that C. S. Lewis cherishes—the determination of a human not to sell her soul for power:

"It is the magician's bargain: ***give up our soul, get power in return***. But once our souls, that is, ourselves, have been given up, the power thus conferred will not belong to us. We shall in fact be the slaves and puppets of that to which we have given our souls[7]" (Lewis 1943).

UM treasures the value of freedom in the same way when the human species is concerned. If the human species has to give up its freedom of the will in order to get safety in return, such species deserves to be gradually replaced by a freer and more moral posthuman species. In that case, it is morally justified to "abolish Man".

References

Agar, N. 2013. Why is it possible to enhance moral status and why doing so is wrong. *Journal of Medical Ethics* 39 (2): 67–74.
Kant, I. 1907. *Religion innerahalb der Grenzen der bloßen Vernunft* (*Religion Within the Boundaries of Mere Reason*). Ausgabe der Preußischen Akdemie der Wissenschaften (Ak. 6: 3–202).
Lewis, C.S. 1943. *The Abolition of Man*. Oxford: Oxford University Press.
Rakić, V. 2014. Voluntary moral enhancement and the survival-at-any-cost bias. *Journal of Medical Ethics* 40 (4): 246–250.
Rakić, V. 2015. We must create beings with moral standing superior to our own. *Cambridge Quarterly of Health Care Ethics* 24 (1): 58–65.
Rakić, V. 2021. *How to Enhance Morality*. Dordrecht: Springer.

---

[7]Please note the similarities with Dostoevsky's "Legend of the Grand Inquistor" in *The Karamazov Brothers*.

The manufacturer's authorised representative in the EU is Springer Nature Customer Service Centre GmbH, Europaplatz 3, 69115 Heidelberg, Germany. If you have any concerns regarding our products, please contact ProductSafety@springernature.com

Printed and bound by CPI Group (UK) Ltd, Croydon, CR0 4YY

25/03/2026

02078197-0009